GOVERNING THE WILD

Governing the Wild

. . . .

Ecotours of Power

Stephanie Rutherford

University of Minnesota Press
Minneapolis
London

Published by the University of Minnesota Press
111 Third Avenue South, Suite 290
Minneapolis, MN 55401-2520
http://www.upress.umn.edu

Library of Congress Cataloging-in-Publication Data

Rutherford, Stephanie.
Governing the wild : ecotours of power / Stephanie Rutherford.
p. cm.
Includes bibliographical references and index.
ISBN 978-0-8166-7440-4 (hc : alk. paper)
ISBN 978-0-8166-7447-3 (pb : alk. paper)
1. Ecotourism. I. Title.
G156.5.E26R88 2011
910.4—dc23
2011028086

Printed in the United States of America on acid-free paper

The University of Minnesota is an equal-opportunity educator and employer.

18 17 16 15 14 13 12 11   10 9 8 7 6 5 4 3 2 1

to Paul Rutherford, my father,
for his trenchant insight and thoughtful guidance

Contents

Governing Nature

Many Stories, One Discourse

In the Hall of Biodiversity at the American Museum of Natural History in New York is a glass case embedded in the floor of the exhibit called "The Crisis Zone." Housed within this case is a number of extinct or endangered animal specimens, the most notable among them being a giant Bengal tiger. Its eyes are alert, round, steady; it regards the viewer almost in surprise at the encounter. Its gaze draws the viewer in, inviting a connection—a mutual recognition of the other and understanding that its wild cousins are under threat. This is charismatic megafauna at its best, and its presence countenances reflection on how things could have gone so badly awry. Its beauty provokes a question: what can we do to stem the tide of environmental destruction so that animals such as this can be saved?

But it is also a dead tiger; no breath will illuminate its luxuriant fur, no life can be detected in its static gaze. And so, while it appears that this tiger offers the museum visitor something, it is, in fact, a one-sided story. It has been killed, stuffed, and displayed by museum curators, first to show people nature from a distant land, and now as a cautionary tale of the excesses of modernity. In either iteration, it did not choose its fate. Rather, it is a sacrificial specimen, its life having been taken in the pursuit of science; it has died so it can never die. And its presence acts as an allegory in a modern bestiary, a morality tale about the impending death of nature. This ossuary of biotic life instructs the visitor about what will happen if we do not act for nature.

The American Museum of Natural History's Hall of Biodiversity is but one of the case studies that animates this book. The others are Disney's Animal Kingdom, an ecotour to Yellowstone and Grand Teton National Parks, and Al Gore and his film *An Inconvenient Truth*. Given the diversity of the cases represented, it would seem that scene described above would be an isolated one, particular only to museums and their visual logic of display and representation. However, the narrative of the tiger does not stand alone. Indeed, part of the task of this book is to chart the ways in which its cases find common cause. So, at Disney's Animal Kingdom, instead of finding a tiger, the visitor encounters a talking recycling receptacle, exhorting in Uncle Sam–like fashion: "I want *YOU* to recycle." The talking recycling bin, in all its frivolity and surreality, actually does serious work. It names an environmental problem and its solution (recycling) without ever having to tackle the thorny issues of what creates excessive waste, or how theme park culture might be implicated in that practice.

In Yellowstone and Grand Teton, it can appear as if nature has already been saved. Visitors can pretend that the clock has been turned back, that nature exists here as it did before westward expansion. Due to the efforts of National Park staff, combined with the foresight of those who crafted the park ideal, visitors can leave their urban homes and encounter an "authentic" and "pristine" nature. The management of the lives of these animals, from reintroductions to tracking, has made it so that they can thrive in this small yet breathtaking parcel of the United States, leaving much of the rest wide open to business as usual. It is in this oasis—this recreated Eden—that the wholesale slaughter of bison, wolves, and bears that made the nation can be forgotten. To stand here is to make modernity's effect on environmental destruction seem distant and the ability to remedy it through more management seem all too possible.

And what might the film *An Inconvenient Truth* have to do with the aforementioned examples? Among the many striking images in Al Gore's 2006 documentary is a time-lapse visualization of a world flooded by rising sea levels. Kolkata, lower Manhattan, and large parts of the Netherlands along with other sites are pictured under water, caused by the impending melt of Greenland and the Antarctic ice shelves due to anthropogenic climate change. While there is vigorous debate about whether these projections are true, the images are indisputably haunting, providing a visual parable of the capacity of humans to destroy our habitat. And yet, in this film, the same civilization that has despoiled nature is simultaneously recuperated. *An Inconvenient Truth* operates through the generation of fear but also redemption: after scaring viewers into somewhere north of panic, Gore reassures us that we already have the tools to forestall the impending environmental apocalypse—we need simply to implement them. Change your thermostat, use energy efficient light bulbs and appliances, buy a hybrid car, take mass transit, switch to renewable energy, plant trees: like the other cases in *Governing the Wild,* the options on the path to environmental salvation are not outside our reach, but what we need is proper management.

What these examples gesture toward is the way that sites of popular culture and environmental education generate scripts that define, circumscribe, and limit how nature is understood. The stories they narrate, while occurring in disparate sites, illuminate the way that various authorities work to tell the "truth" about nature. Nature threatened, managed, and recuperated are the dominant themes. In all four cases, it appears that nature exists for, is transformed by, and can be saved through human action. In each story, nature is enrolled in a capitalist regime, where it is consumed, both through the senses and as commodity. All four examples point to the surgical removal of the teeth of environmentalism, excising the radical thrust of critique and replacing it with a path forward that sidesteps deeper reasons for environmental problems. So, I suggest that much more goes on at these sites than the mere experience of a preexisting nature. Rather, these sites create nature in particular ways, but simultaneously erase the fact of its cultural production. And I would argue that the result of these kinds of tales is that other ways of encountering nature are rendered unthinkable, other stories unsayable.

Governmentality

But while each of the cases shares similarities, a main argument of *Governing the Wild* is that they also offer important differences that allow for a reconceptualization of what it means to think about and manage nature. I contend that what occurs at these sites is a form of *governing,* defining the imaginings, discourse, and practices that make up what the Western bourgeois subject comes to know as nature. But each case offers a specific kind of governance, and in combination they demonstrate the myriad ways in which culture can work as a site of relations and contestation of power.

In sketching out this argument, I rely on Michel Foucault and his interlocutors. More specifically, I utilize the analytical scaffolding provided by his writings on governmentality. One of Foucault's most lasting contributions to the intellectual landscape has been his analysis of power. In a counterintuitive move, Foucault decries the traditional understanding of power as only domination and repression. Instead, he argues that although power does discipline and control, one of its central features is its productiveness: "In fact, power produces; it produces reality; it produces domains of objects and rituals of truth" (Foucault 1995, 194). But, in its production of knowledge, truth, and subjectivities, Foucault (1990) is at pains to elucidate that power is not possessed or held, but rather circulates via networks that work through and produce different bodies, discourses, institutions, and practices. Foucault's work on governmentality offers one iteration of his larger understanding of the operation of power in the modern world, one that in particular sought to understand liberalism as a form of political rationality. Foucault suggests that before the sixteenth century, power was rooted in the sovereign; in particular, the king's juridical right to oversee property and subjects, an authority often exercised through the public execution (see 1995). However, in the 1500s, a series of transformations began, including the rise of the Protestant Reformation and a sense of the utility of Christian pastoral power, the debate about the reason of the state, and emergent theories on self-government, that initiated different questions about the nature of governing itself: "How to govern oneself, how to be governed, how to govern others, by whom the people will accept being governed, how to become the best possible governor—all these problems, in their multiplicity and intensity, seem to me to be characteristic of the sixteenth century" (Foucault 1991, 87). Building over the course of two centuries, the refashioning of the idea of government was bolstered by the demographic boom of the eighteenth century, which reinforced the need to manage the population itself. Demographics, statistics, and insurance established new power/knowledge formations that organized individuals into populations whose characteristics could be measured, assessed, and managed. The individual and population increasingly became objects of knowledge for government, where the aim is to strengthen the state through the exercise of tactics and the construction of knowledge rather than the imposition of law (Foucault 1991, 2003a). Governing becomes the construction of certain truths and their circulation via normalizing and disciplining techniques, methods, discourses, and practices that can be rooted in the state, but also extend beyond it to stretch across the social body (Foucault 1990). In its broadest articulation, governmentality is made up of "techniques and procedures for directing human behavior. Government of children, government of souls and consciences, government of a household, of a state, or of oneself" (1997, 82). However, this does not signal the death of sovereign power. Rather, what we see in this new modality of governing is a triad of sovereignty, discipline, and government, where each is reformed "within the concern for the population and its optimization (in terms of wealth, health, happiness, prosperity, efficiency), and the forms of knowledge and technical means appropriate to it" (Dean 1999, 20).

One key aspect of this transition to the "art of government" is the emergence of biopolitics. Unlike his earlier work in *Discipline and Punish* (1995; orig. pub. 1975), where he concerns himself with the construction of docile bodies, it is in Foucault's later work in *The History of Sexuality,* volume 1 (1990; orig. pub. 1976), and more fully in his lectures at the Collège de France (2003a, 2007, 2010) that he turned his attention to biopolitics. In contrast to disciplinary power, biopower takes root through the regulatory controls of the population (rather than only the individual) in the management of life—birth rates, life expectancy, health, and well-being—which act as indicators of the population that began to increasingly matter to those who govern. Indeed, Foucault contends that biopower signals the "entry of life into history," or as Rabinow puts it, biopolitics "brought life and its mechanisms into the realm of explicit calculations and made knowledge-power an agent of the transformation of human life" (1984, 17). More succinctly, biopolitics is "the politics of life itself" (Rose 2006). However, it is important to recognize that even though there is this distinction between disciplinary power and biopower, they work in tandem: one individualizing and the other collectivizing, governing the conduct of each and all.

Part of this biopolitical regulation works through the construction of the self and the ways one is situated within the broader population. Here, the allied notions of subjection and technologies of the self are important. Subjection is the process by which the conditions and parameters are set for individuals to "tell the truth about ourselves"; those discourses and practices that allow us to think of ourselves as autonomous, self-directing individuals who make choices and act upon the world (McHoul and Grace 1993). As Foucault tells us, "[p]ower is exercised only over free subjects, and only so far as they are free" (1982, 221). For power to operate there must be the freedom to be otherwise, to choose different subject positions out of a range of possibilities. It is through the shaping of different options—different subjectivities—that we come to know ourselves. Subjection is thus simultaneously individualizing and totalizing, asking the subject who she or he is individually, but also where he or she fits in the totality (Foucault 1982). It occurs through the techniques and practices that make a particular subject, and concurrently becomes one unit through which subjects become part of the population that can be governed and normalized for the good of the totality—whether that totality takes the form of the nation, the biome, or the imagined community of the global environment (Foucault 1994, 1995). But this is not to say that the making of subjectivities is something that is done to or imposed upon individuals. As Dean remarks: "Regimes of government do not determine forms of subjectivity. They elicit, promote, facilitate, foster and attribute various capacities, qualities and statuses to particular agents" (1999, 32). Subject formation is a productive process; subjectivities are always *becoming*. Thus, as Foucault reminds us, this is an active engagement, where "individuals are the vehicles of power, not its points of application" (1980, 98).

One part of the making of certain subjectivities is through technologies of the self. These are the ways in which people choose to become certain kinds—often more *virtuous* kinds—of subjects, through the application of techniques for improvement. As Rose has noted of the neoliberal subject, there is a call to "[b]ecome whole, become what you want, become yourself: the individual is to become, as it were, an entrepreneur of itself, seeking to maximize its own powers, its own happiness, its own quality of life, through enhancing its autonomy and then instrumentalizing its autonomous choices in the service of its life-style" (1996, 158). This works in very particular ways, but at its base is the notion of incompleteness: that there is something else we need to be, do, or have to become a coherent and self-actualized subject (Miller 1993). And in large part it functions through the kind of pastoral power in a quasi-therapeutic relationship, where the expert gives kindly advice to the subject in the journey that strives for an unattainable completion. It is, in essence, self-fashioning: through the examination of conscience, the process of confession to an expert and the renunciation (or ascetic rejection) of behavior that might be understood as "abnormal," the subject's conduct can be corrected and the individual can embark on her or his path to self-fulfillment. However, it is important to remember that there are erasures and foreclosures in the way people can conceive of themselves, and the performance of different subjectivities are read as (un)intelligible differently, something to which Foucault paid significantly less attention. This will be explored further in the subsequent chapters.

What makes this approach different from other studies of rule? What new insights does it provide that are, for example, missing in Gramscian (1992) hegemony theory or Putnam's (1993) theories of governance? Is it simply another esoteric label to talk about the same old thing? What differentiates governmentality as an analytic of power from either Gramsci or Putnam is its approach to power. First, it seeks to take apart the self-evidence or truth of governing, revealing its historical contingency and entrenchment in the social, political, economic, and cultural (and I would add biophysical) contexts that produce it. Further, it decenters the state as seat of power: power bleeds across the social body in such a way that governing occurs in multiple sites and through a myriad of techniques. And finally it asks the "how" questions that Dean emphasizes:

> This approach thus stands in contrast to theories of government that ask "who rules?," "what is the source of that rule?" and "what is the basis of its legitimacy?" An analytics of government brackets out such questions not merely because they are stale, tiresome, unproductive and repetitive. It does so because it wants to understand how different locales are constituted as authoritative and powerful, how different agents are assembled with specific powers, and how different domains are constituted as governable and administrable. (1999, 29)

The strength of this sort of interrogation is that it allows for different kinds of assertions about the ways in which modern rule operates, ones that are central to the cases presented in this book. First, governmentality articulates the possibility that power is exercised in multiple sites, through different discourses, and often outside the traditional boundaries of the state. For if the state is no longer the automatic seat of government, how are people governed in alternative sites? How do these sites become authorized? Through what techniques and practices? How does one examine the microphysics of power? How does the constitution of the self contribute to governmentality? These are questions that I think take note of the state but also go beyond it, bringing nongovernmental organizations, corporations, research institutions, the media, and other actors into the challenge of tracing modern rule.

Although the analytic provided by governmentality is a fruitful one, that does not mean it is without problems. I will outline related issues here, which I will both expand upon and seek to remedy through the course of *Governing the Wild*. First, Foucault and studies that use governmentality often seem to marginalize or erase difference in the administration of rule. But there are important things to be said about the ways that governmentality works through difference, creating alliances as well as divisions among people.[1] There are possibilities and foreclosures in the way people can conceive of themselves, and the performance of different subjectivities are read as intelligible differently. This is particularly true with reference to race, class, and gender; not all subjectivities are equally available to all. For example, it is important to note that certain people are called upon to fulfill the role of those who "care" about the environment and its protection, as it is remade into a new space of governmentality. And it is most often women, the poor, and racialized people who are excluded, or sharply defined, within this regime (Luke 2003). How are different bodies governed in different ways based on social markers like race, class, gender, and sexuality? How does a microphysics of power target people in distinct ways based on difference? How do bodies that are marked by difference encounter spaces of governmentality differently? How might they read these spaces in complex and unpredictable ways? These are the kinds of questions I also want to consider in this study.

Similarly, studies that have employed governmentality as a heuristic sometimes intimate that the techniques, practices, and policies they describe operate and circulate as intended with very little deviation. This ignores agency (both human and nonhuman) as well as the interstitial slippages that can occur in the application of power and the moments of instability that emerge as a result. Because, of course, the construction and performance of rule is always the result of contested engagement (O'Malley, Weir, and Shearing 1997). Governing does not arise as a fully realized project, but is continuously in need of re-articulation and revision. Thus, *Governing the Wild* illuminates the nodes of power/knowledge that shape how we come to understand things as truth, while recognizing that this truth and its government is always becoming, necessarily uneven, and often contested.

A final critique I would like to assert is that while studies in governmentality pay attention to the operation of power, they are often abstracted to such a degree that it is difficult to see their impact on human and nonhuman life. In particular, the nonhuman is often nowhere to be seen in studies that look at rationalities of rule. This gap is addressed in the next section.

Green Governmentality

The reader might be asking "What does all this talk of political rationality have to do with the Disney's Animal Kingdom or *An Inconvenient Truth*?"

Or perhaps more trenchantly, what does this have to do with nature? It's true that nature was never high on Foucault's list of priorities; in fact, he indicated a definite distaste for it (Éribon in Darier 1999). This might give some indication as to why nature was rarely included in Foucault's analyses.

Yet, Foucault's fixation on the management of populations almost inevitably leads to a consideration of the biophysical world. As Paul Rutherford (1994) reminds us, the government of population must include the very environment from which humanity subsists. Indeed, nature—claims on the land, the construction of wilderness, ideas of human nature, human/non-human interaction—is one area in which the messy politics of discursive construction comes to the fore, making it a particularly interesting site to interrogate the exercise of power (Moore, Kosek, and Pandian 2003). The ways in which the environment is constructed as in crisis, how knowledge about it is formed, and who then is authorized to save it are important for understanding how the truth of nature is produced and governed.

In recent years, scholars have taken up this gap in Foucault's work, arguing that we need to think about a form of governmentality specific to environmental relations, from forest and groundwater management to environmental nonprofits and green consumerism.[2] Within green governmentality frameworks, the environment is understood not only as a biophysical reality, but also as a site of power, where truths are made, circulated, and remade. Green governmentality allows for the understanding of nature as artifact, where knowledges and subjectivities are made in and through discursive and nondiscursive practice. This is especially important because, of course, the saving of nature is often taken for granted as an innocent endeavor, a noble exercise for the good of all life. However, mainstream environmental narratives are usually underpinned by "the one-world discourse," an assertion whose basis is the notion that we are all connected through our intertwined ecological fate (Y. King 1997, 1). Thus, what emerges is a "dominant storyline of the 'fragile earth' under stress from human action and in need of care and protection from an imagined global community" (Macnaghten 2003, 65). This way of producing the environment and its resources as bounded elicits the discourse of the limits of the earth, a central tenet of environmental politics. The production of this kind of truth about nature necessitates its regulation and management by particularly located experts. Within this discursive regime, we must situate various state-based and international agents, corporations, and the earth sciences. But one cannot leave out environmental organizations, which also shape the truth about nature, and seek to regulate and ameliorate its (ab)use. Indeed, there are now more than a hundred thousand environmental organizations worldwide working to regulate and manage nature (Emel 2002), or at least there were in 2002 when Emel penned her piece; one imagines the number has only grown. Further, we see the emergence of corporations acting as environmental funding agencies, as initiatives like Disney's Wildlife Conservation Fund attest, something that will be taken up again in chapter 2. The result of both the number of organizations and their broadened political function has meant that "most environmentalist movements now operate as a basic manifestation of governmentality" (Luke 1999, 121).

The Commodification of Nature

This study seeks to extend these important insights further, however, not only talking about discourses, practices, subject formation, and fields of power, but also about profit and materiality. So, for example, sometimes it is precisely through the definition of nature as a space of regulation that it also becomes consumable in very particular ways and by very particular people. To understand the ways in which sites of consumption are important for how nature is imagined, deployed, and ruled, there must be traffic between notions of governmentality and commodification. First, notions of commodification point to the ways in which nature is enrolled in the service of capital, which is a key function of its governing. Further, it emphasizes the material reality of rule; animals, plants, people, places, institutions, and corporations all shape and are shaped by the discourses that make up what gets defined as nature. These impacts are not simply limited to the realm of ideas but produce material costs and benefits. In other words, discourse does not leave the land untouched: it makes and remakes nature for consumption by particular people at specific times. At the same time, materiality matters. The biophysical world writes back, as it were, shaping how discourses can be imagined and deployed. Moreover, studies of the commodification of nature have charted how the consumption of nature is a highly gendered, racialized, and classed project (see for example Escobar 1996; Katz 1998). Thus, a main argument of this book, which I only alert the reader to here but flesh out more fully in the subsequent chapters, is this interplay between these two theoretical frames, making the case for considering how governmentality and commodification work in tandem to produce the truth about nature as not only scientized, knowable, and measurable, but also consumable.

To say that nature is commodified is not saying anything terribly new. The exploitation of natural resources has been a defining feature of capitalism since at least the colonial contest and, indeed, nature as modified by capital has served as the main export and nation-building narrative of the United States, Canada, and other "white settler" countries. However, it is worth remarking that this trend has accelerated and now has come to encompass all manner of things that were once thought as outside of capital. The buying, patenting, and selling of biodiversity, the bio-tech revolution, the marketing of botulism for cosmetic surgery in the form of Botox, the notion of growing human organs for transplantation: these all speak to a new and increasing kind of commodification that goes far beyond resource extraction and enters the domain of life itself. As Rose (2006) remarks, it is not so much that these scientific efforts are preexisting and then refashioned as a marketing strategy, but rather that the biosciences have been reimagined as a mainstay of capitalism. Rather than simply treating nature as an externality or limitless source of raw materials, Escobar (1996) argues that capitalism has adopted a conservationist focus, maintaining biodiversity so its genetic harvest can be reaped. Yet another wrinkle to this conservationist turn is that environmentalist scripts have themselves increasingly become the subject of commodification. The rise of environmentally friendly consumerism has led to an explosion of ecotour operators, organic agriculture, and companies that refashion their products as part of this growing movement. In each of these cases, nature is remade for a consumption that one is supposed to feel good about; this is the kind of consumption that, purportedly, can change the world. Viewing these nature narratives as not only about power but also about money—which can rarely be separated—reveals a new dimension in stories about nature. What emerges, then, is both a political economy and cultural politics of nature, which is thoroughly enmeshed in the entanglements of power, culture, nature, and profit that define our world.

Different Kinds of Green Governmentality

Governing the Wild seeks to expand and complicate the notions of green governmentality outlined above through its four case studies. At first glance, each of these sites seems disparate, but I contend they share much in common. First, they are all emblematic sites, which carry a resonance beyond their particular localities or operate as global phenomena. The American Museum is known as a great repository of specimens of, and knowledge about, natural history. It enjoys preeminence in the realm of academic and popular culture. Disney is one of the cultural producers of childhood, crafting and recrafting the images, ideas, and sometimes values that are carried on into adulthood and freighted with nostalgia. Yellowstone is a veritable nostalgia machine, harkening back to the halcyon days of western expansion in the United States and serving as a monument to nationalism, the frontier, and wilderness. A visit to each of these represents a pilgrimage to a sacred part of American culture. *An Inconvenient Truth* functions in a similar vein, offering a compelling narrative of greed and ingenuity, nature and technology. Each case study has a currency that extends beyond the country that houses them; they have become iconic in a panoply of places or artifacts that have achieved global significance.

But these are not only spaces of culture. They are also sites of nature, particularly defined and deployed. What I want to suggest is that each of these sites examined in this book offer *different manifestations* of green governmentality, operating on registers that Foucault did not anticipate. This assertion is further elaborated through the chapters, but I will give the reader a flavor or what is to come here.

The Hall of Biodiversity at the American Museum of Natural History offers an excellent example of scientific dimensions of green governmentality. This site shows how science, historically and in the contemporary moment, molds what nature is and can be. The Hall of Biodiversity exhibit acts as an agent in constructing a certain kind of nature: a scientized assessment of global nature under threat. The characteristic way of seeing provided by the American Museum is bolstered by its legitimacy and authority; science is already understood as a vehicle for truth-telling. I contend that the Hall of Biodiversity acts as an inventory of nature, an enframing of biodiversity in its totality through the language of science. Through these mechanisms, the Hall of Biodiversity provides characteristic ways of seeing nature, maps a biopolitical present and future, generates technologies for its proper management, and offers particular subject–positions for visitors to take up. Thus, relying on the explanatory power of science and the tactics of display characteristic of museums, the Hall of Biodiversity is a space of power through which visitors come to know an objective, dispassionate, and classified understanding of the natural world, eliding both the fleshy lived realities of nature and the other ways that one might encounter it.

By contrast, Disney's Animal Kingdom offers the reader a window into corporate green governmentality. The Animal Kingdom demonstrates that a corporation can successfully organize a biopolitical project, and because of this, function as an agent of green governmentality, all the while generating profit by deploying nature and eliciting emotional reactions to it. The Animal Kingdom presents nature as simultaneously fantastic and spectacular while donning the mantle of legitimacy by positioning itself as nature expert and reimagining itself into a kind of nongovernmental organization. The theme park is a sensory manifestation of a corporate effort to articulate and shape a particular view of animal life and, in doing so, generates not only profit, but also knowledge, pleasure, and specific subject positions. Interestingly, Disney works to produce a discourse of "un-truth": melding fact and fiction in typical Disney fashion to produce a version of the "truth" of nature, or at least what it should be. Disney uses fantasy and morality to craft particular subjectivities steeped in the ethos of consumption yet satisfying an ethical imperative, making them seductive to embrace.

The ecotour to Yellowstone and Grand Teton provides a roadmap for the government of what we come to know as natural beauty—aesthetic governmentality. In this way, it operates as an intellectual factory producing the visual grammars of "pristine" national natures. Through the use of discourses like science and conservation, and technologies like the gearfetish and photography, a particular commodified version of nature was offered and accepted. Each day of the ecotour highlights what vistas count as beautiful and hence reproducible, what animals need to be seen, and how it all may be represented in the assemblage of lens/eye/animal. As such, there is a normal way of encountering nature: away from home, singular in its grandeur, devoid of humans, and aesthetic in composition. That this tour takes place in two national parks only serves to reinforce its connection to other structuring narratives of nationalism, wilderness, and nature that shore up its role as an agent of green governmentality.

And finally, Al Gore and his film *An Inconvenient Truth* put forward a kind of moral governance on the subject of climate change. Like the museum, the Animal Kingdom, and the ecotour, I assert that *An Inconvenient Truth* relies on the impartiality and unquestioned truth of science to warn of impending global apocalypse, reinscribing the preeminence of this way of understanding nonhuman nature. But more than its appeal to the scientific endeavor, I contend that this film also works through narrative—through the affective potential of storytelling. Through nostalgia, fear, and nationalism, as well as the dispassion of science, Gore offers viewers a moral green subjectivity through which to encounter the natural world. He acts as Foucault's truth-teller; an agent authorized to name a crisis and propose solutions before we all end up under water. Relying on affect and knowledge—heart and mind—Gore builds the perfect moral green citizen: one that doesn't consume less, but consumes appropriately. In this way, Gore inserts an ethical dimension to the practice and politics of green governmentality.

Thus, each of the sites examined offer unique case studies that suggest deeper insights into the particular workings of a specific kinds of green governmentality, lending empirical specificity and depth to notions of governmentality. Of course, these distinctions briefly outlined here are too discrete—for example, all the cases could be said to employ science, profit, aesthetics, and ethics in complex and sometimes unpredictable hybrid combinations. Moreover, to greater and lesser degrees, all deploy simulacra, fantasy, and play to weave a seductive narrative of nature. And each space dictates forms of behavior, practice, and politics that they assure us will forestall environmental calamity. In this way, these sites find common cause, an interesting notion when one considers the discursive space between a museum and a theme park, for example. However, even in their similitude, they still offer manifestly different kinds of green governmentality. An examination of each not only expands the range of possibilities when thinking about how governing works with relation to both the human and nonhuman, but also offers the potential for considering what a space outside of governmentality might look like.

A Note on Methodology

Thomas King, First Nations novelist and academic, tells us "the truth about stories is that's all we are" (2003, 2). He contends that "[s]tories are wondrous things. And they are dangerous" (9). If King is right, then the stories in *Governing the Wild*, which range from cooing over bison to global environmental apocalypse, have an important place in the world. They provide us with the tools to understand nature, to experience it, and to rescue it from destruction. The purpose of this book, then, is to query these stories, to ask what the consequence is of their telling. In this way, *Governing the Wild* is all about stories, how they function, to what effect, and how we might tell them in different and more just ways.

But this book is also about more than stories. More properly, it is about how narratives—that cluster into discourses—have material dimensions and effects. In some articulations, discourse is linked only to language, conceived of as a regulated linguistic space of utterances and statements about some aspect of the world. This is, of course, important because talk is one of the key ways we construct reality. However, it seems to me that this description of discourse is somewhat impoverished. Communication is not only about language, or what we say, don't say, or can't say. It is also about those actions, sites of production, practices, embodiments, and images that support or resist a particular way of thinking and talking about a subject. For example, the discourse of global environmental problems includes not only the statements made about, say, climate change, but also environmental practices that support such articulations, the legal regimes that emerge to regulate it, the graphs of greenhouse gas emissions or the images of ice shelves melting, the beings like polar bears that become enrolled in this regime, and the scientific and political practices that produce it. And so, discourse for me is the ensemble of sign systems (linguistic, practical, visual, and embodied) that construct a particular topic and has truth-effects in the world.

Discourse, then, is necessarily about power. But it is important to try to distinguish discourse from ideology. Ideology, at least in the Marxist tradition, is imagined as an oppressive mechanism that seeks to coerce or obtain consent for a particular agenda. In this iteration, the exercise of power seems to come only from above—an authority that aims to control. However, Foucault's articulation of power as not only repressive but inherently productive provides a key differentiator between ideology and discourse (Mills 2003). For Foucault, power, and thus the ability to craft particular discourses, is not possessed or held but circulates via networks that work through and produce bodies, subjects, discourses, practices, institutions, and representations. This can come from anywhere, not necessarily the halls of power. Moreover, unlike ideology, within discourse itself there is a possible space for resistance. As Foucault states in *The History of Sexuality:* "[D]iscourses are not once and for all subservient to power or raised up against it, any more than silences are . . . discourse transmits and produces power; it reinforces it, but also undermines it and exposes it, renders it fragile and makes it possible to thwart it" (1990, 100–101). Thus, discourse, in contrast to ideology, comes from multiple sites and is not a finished project, but is continuously renegotiated, re-articulated, and resisted.

With this definition in mind, I employed discourse analysis as my primary methodological frame to examine the cases in this book (Fairclough 2003; Wetherell, Taylor, and Yates 2001). I treated each place and the film as visual text to determine what the underlying narratives were in each one, examining the text, images, and spectacles provided for deeper insight into the stories they told. Moreover, in each site I was a participant observer, acting as a museum visitor, theme park guest, ecotour client, or filmgoer, respectively. I noted what people did in these sites, what they said, and how they interacted with various objects and ideas on display. I accompanied this fieldwork with a close reading of the range of Web sites, images, press releases, educational material, films, advertisements, statistics, and archival material available on and at each of these places I visited as well as the film *An Inconvenient Truth*. Further, I conducted interviews with key informants at each of these sites, with the exception of the Animal Kingdom (see chapter 2 for more detail). At the American Museum, I interviewed the lead curator for the Hall of Biodiversity and two staff members at the Center for Biodiversity and Conservation. On the ecotour, I interviewed our guide and our logistical organizer. The purpose of these interviews was not to perform an ethnographic analysis but rather to access background information not available to the public about each case study.

Something must be said, however, about what might appear to be divergences in the application of this methodological frame, or what might be termed as a lack of parallel construction of the cases. For while I want to suggest that there are strong similarities between the cases, it is true that each of the sites examined in *Governing the Wild* present different kinds of experiences. As such, I conducted observations at each site, but sometimes in some chapters this is more clearly participant-based than others. For example, at Disney's Animal Kingdom and throughout the ecotour, I was a direct participant, engaging with other guests/tourists, experiencing the same phenomena. My reading of *An Inconvenient Truth* and the persona of Al Gore was a more solitary endeavor, involving an analysis of the documentary, of course, but also its related books, speeches, and media artifacts. This is a fundamentally different kind of discursive approach than those based on a sensory encounter of a place. The American Museum is a kind of bridge between these two poles, offering both a physical experience while emphasizing the personal nature of exhibit consumption, not conducted in tours, but through watching others consume its spectacle. Because of these dissimilarities, the chapters read differently. But what *Governing the Wild* aims to do is bring together disparate kinds of nature–culture practices—some emphasizing the visual, others affectivity, some more intellectual, others playful—to think about how such places might coalesce into an emergent, though often contingent, regime of truth. The broad banner of discourse analysis allows for this kind of empirical promiscuity.

However, the practice of discourse analysis itself represents a contested terrain. As Terry Threadgold points out, "This dissemination of *the idea* of discourse analysis 'within a political multiplicity' which indeed has no single mother tongue . . . or any single clearly defined approach to the subject, is what makes it so *difficult* for those who would use it to know where to start or even to be sure what it is" (2000, 40–41, emphasis in original). Loretta Lees (2004) has potentially made this quagmire somewhat less sticky by suggesting that there are two forms of discourse analysis: a Marxist-inspired approach that seeks to understand the operation of ideology and hegemony, and a Foucauldian method that is concerned with the production of regimes of truth and the process of subject formation. The aim in this second strain, then, is not to determine whether discourses are true or false, or to explore preexisting identities, as is the case with the first iteration of discourse analysis, for, to paraphrase Foucault, this would be chimerical. Rather this second vein of discourse analysis seeks to understand how discourses operate and to what effect. As Poynton and Lee have indicated, discourse analysis is an effort "to capture the regularities of meaning" (2000, 6) so as to understand how the circulation and iteration of particular systems of language and practice come to be known as "truth."

However, the distinction that Lees provides is, for my purposes, too rigid. For while my study primarily falls within Lees's second articulation of discourse analysis, my view is also that that there are better stories to tell, better ways of thinking about the relationships between humans and nonhumans. Indeed, I agree with Haraway when she writes that "it is a rule for me not to turn a dissolving eye onto straw problems, not to 'deconstruct' that to which I am not emotionally and politically vulnerable" (1996, 348). The stories told about the relationships between humans and nonhumans matter to me. Research, in my view, is necessarily political: it opens up a space of critique to imagine the world differently. So, to be sure, I employ discourse analysis as a means to understand how discourses of nature are formed, circulated, and taken up. I am interested in disrupting

what is taken for granted as "truth," and reading power through its establishment as self-evident, laying bare how particular discourses, institutions,
and practices have emerged through specific and historically situated systems of rule. But I am also keen to reveal those knowledges that have been
foreclosed through the effects of the exercise of power. As such, I turn an
eye to the question of un-writing or rewriting the well-worn groves of
these stories. If we can understand how discourses about nature produce
particular ways of understanding the world, then part of my practice of
discourse analysis opens up the possibility to think otherwise. This will be
taken up more fully in the conclusion to *Governing the Wild*.

CHAPTER 1

Ordering Nature at the American Museum of Natural History

WHEN YOU ENTER the American Museum of Natural History in New York from the 77th Street entrance, you encounter a profoundly striking diorama depicting the Dutch colonization of New Amsterdam (now Manhattan). A group of men are foregrounded, two Indigenous and two European. The Europeans, one brandishing a rifle and the other with a hand outstretched, invite the Indigenous men (smaller in size) to approach in supplication and bestow the corn and jewelry they carry, presumably in the form of a gift to the newly minted American settlers. The Indigenous men are shown wearing loincloths and feathers—the quintessential body of the "primitive" marked by its "naturalness." The diorama is spatially halved, the one side reflecting European mastery of the sea and technology, the other depicting the Native as outside the thrall of modernity. But the ships in the far background herald further settlement of the area: the inevitable march of civilization. This is the story of a sanitized empire. Erased from this diorama is the actuality of colonialism—nothing short of genocide. Instead, we are asked to imagine this diorama as a value-free depiction of contact and exchange between the Indigenous people who called that land home and the seemingly benevolent European settlers of the New York area.

By now, this is a familiar—if hotly contested—narrative of colonization. It has become easy for us to deconstruct this diorama thanks to the work of Haraway (1989), McClintock (1995), T. Mitchell (1988), Said (1979), and others to reveal how it tells the stories about race, gender, nature, and nation that lend it coherence. This, of course, is not an uncommon representation, and the story it tells us serves as a constant trope in the Western imagination of the "other." Indeed, it is a very convenient retelling of the history of the United States (and "white settler" countries more generally), which reinscribes Native people as close to nature—hence in need of improvement—and positions the European settlers as the genuine inheritors of the emerging nation. This diorama speaks without speaking of the ways in which power and oppression are naturalized in the construction of nations.

This is the kind of narrative that I anticipated finding at this research site: the Hall of Biodiversity at the American Museum of Natural History (American Museum). I imagined I would be presented with a straightforward story line that legitimizes the imperial impulse under the auspices of saving global nature. But I was wrong. After completing my research at the Hall of Biodiversity, I came to understand that the governing of nature is at times a subtler and more complicated affair in this exhibit, simultaneously reinforcing and challenging some of the key narratives about nature. Indeed, the exhibit is inflected with good intentions and in many ways offers one of the more complex treatments of the entanglements of nature and culture in the making of environmental problems than the other cases presented in *Governing the Wild*. But I think what the American Museum does through this exhibit is offer an example of governmentality, where particular kinds of science operate as truth-telling mechanisms to construct how nature is understood. So, in examining this site, I attempt to answer the question of how the exhibit works to produce a certain form of "truth" about nature by exploring both what is memorialized and, in the act of remembering, what is forgotten. I also pay attention to what is made legitimate for display and how it is organized, as well as how the visitor's gaze is constructed. Succinctly, I am interested in how power and education intersect in the Hall of Biodiversity.

In examining the site with these questions in mind, I came to realize that the Hall of Biodiversity attempts nothing less than an inventory of nature—a complex representation of biodiversity in its totality. It tells the "truth" of the world through science and display, providing characteristic ways of seeing nature, generating biopolitical maps of biodiversity, reinforcing particular narratives of nature and culture, and constructing green subjectivities. By defining what nature is, how it is under threat, why it is important, and how we can mediate its crisis, this exhibition seeks to enframe nature and redeploy it for effective management. In this way, it operates as a producer, instrument, and means for the circulation of particular kind of green governmentality.

The History of Natural History

Natural history as a systematized discipline emerged in the seventeenth and eighteenth centuries during a period that Foucault (2004) aptly christened "the age of the catalogue." To be sure, however, collections of curiosities existed before this time. As Hooper-Greenhill (2001) has described, collections of objects, which might be understood as proto-museums, were common in sixteenth-century Europe. Of these collections, the German *Wunderkammer* (wonder rooms or cabinet of wonders) is perhaps both the most famous and representative. This "irrational cabinet" of the Renaissance period displayed an assemblage of objects that, according to Hooper-Greenhill, reflected collectors' attempts to give voice to their view of the world. And so, the cabinets or rooms did possess a kind of rationality, although this rationality did not operate through the categories of display that have now become familiar in modern museums. Instead, the boundaries between nature and culture, history and science, fact and fiction, were both permeable and changeable, and the hidden relationships between seemingly disparate items could be explored. Thus, cabinets of curiosity contained all manner of spectacular and exotic items: objets d'art and paintings, real and fabricated natural history specimens (like unicorn horns or dragon scales), "monsters" and oddities, relics and archaeological artifacts, all meant to demonstrate the worldview, as well as knowledge and power, of the collector.

If the Renaissance was characterized by both the display of spectacular strangeness and the attempt to find resemblance between objects, then the classical *episteme* looked upon such Renaissance rationalities as "muddled, confused, and disorderly" (Hooper-Greenhill 2001, 134).[1] Hooper-Greenhill notes that the voyages of discovery, combined with new experimental methods of the seventeenth century, made Renaissance ways of knowing insupportable. The result was one of Foucault's characteristic ruptures between different conditions of possibility for explaining and understanding the world: a shift in the grid of intelligibility. In relation to natural history, he describes this rupture in Latourian terms: as the purification of nature and culture. In Renaissance times, he notes, "all that existed was history" (2004, 140). The borders between nature and culture were much more fluid. As such, histories of plants, for example, included morphology (physical form and structure) and botanical anatomy but also stories and legends that surrounded them. With the coming of the classical *episteme,* this form of knowing becomes displaced, and history and nature are organized into distinct and independent realms. Foucault argues that seeing becomes more focused, based on observable "fact," and jettisoned of anecdote, fantasy, or myth. Knowledge about the natural world, then, becomes defined through an *inattention* to the entanglements of nature and culture that characterized Renaissance knowledge. As Foucault remarks, "the essential difference is what is *missing*" (141, emphasis in original).

With this move to the classical *episteme,* not only is seeing limited but it simultaneously becomes the central mechanism to apprehend the world and its natural mysteries. Through technologies like the microscope, natural historians were able to examine minute aspects of plants, animals, bacteria, and minerals. This allowed for the documentation of morphology to find similarities and differences so that each element of the natural world could be sorted into its proper place. Indeed, Foucault comments, "the blind man in the eighteenth century can perfectly well be a geometrician, but he cannot be a naturalist" (2004, 144–45). As Conn has remarked, natural history "made a fetish of the observable fact" (1998, 33). Through this reliance on vision, which was presumed to be both innocent and truthful, the natural historian sought to render knowable that which already existed in nature: its own orderly structure. So, there were collections of natural items before the seventeenth century. However, it is with the birth of natural history that collections can be named, ordered, and made meaningful to science in a systematic way involving observation and objective analysis.

The notion of revealing nature's order achieved its purest expression through the development of Linnaean taxonomy in the mid-eighteenth century, the grid that made the project of classification thinkable. Taxonomy allowed for the systematic treatment of objects, and with it, "complete enumeration became possible" (Hooper-Greenhill 2001, 134). The global project of seeking out specimens for natural history carried out by Linnaeus, von Humboldt, Buffon, Darwin, and their disciples made this aim manifest. Of course, other systems of taxonomy were present before Linnaeus' travails in the wild. In an oft-quoted passage (see Hooper-Greenhill 2001, 4) from *The Order of Things,* Foucault recalls coming upon a description of a Chinese taxonomic system: "animals are divided into: (a) belonging to the Emperor, (b) embalmed, (c) tame, (d) suckling pigs, (e) sirens, (f) fabulous, (g) stray dogs, (h) included in the present classification, (i) frenzied, (j) innumerable, (k) drawn with a very fine camelhair brush, (1) *et cetera,* (m) having just broken the water pitcher, (n) that from a long way off look like flies" (2004, xvi). This kind of taxonomy, so eccentric and comical to the classical and modern mindsets, jarred Foucault into the understanding that if there were different ways of ordering the world, then perhaps the current method was a contingent product of power and history rather than nature revealing itself to the observing eyes of the scientist. Foucault has been critiqued for just this kind of analysis. Conn (1998, 12), in arguing against the pervasiveness of Foucauldian approaches to the museum, states: "The ways in which museums created, reified, and institutionalized categories of knowledge as unassailable truth does indeed reflect the operation of power. . . . But having said that, we must also acknowledge that understanding cannot happen and meaning cannot be constructed without some set of categories. Further, this identification of categories of knowledge with power risks ignoring the necessary intellectual function that classification plays." While I think it is true that some process of classification is important, for if confronted with the totality of biodiversity one might tremble in the face of its vastness, I think that Conn's comments actually serve to reinforce the utility of a Foucauldian analysis. For the question then becomes, whose categories count; whose truth matters? Clearly not those of the Chinese encyclopedia. It is no accident that particular versions of classification have the effect of "truth," while others appear to be unsupportable. Thus, natural history both obscures and lays bare, producing the very objects that it claims only to catalogue. In this way, natural history and its classification is a knowledge/power nexus.

To sum up, the history of natural history is one that moved from the spectacular and often fictional display of fantastic objects to the collection, naming, classification, and arrangement of objects in a way that revealed nature's plan. In the field of natural history, then, from Linnaeus and Darwin but even in present day cladistics and genetic classification,[2] the world is remade—rather than being incomprehensible or containing elements beyond human understanding, nature becomes orderly, classifiable, and knowable.

Museums as Spaces of Governmentality

When natural history (as well as art and artifacts) moved from the private rooms of kings and was opened through public museums, the way these collections functioned also changed. Before the advent of the museum, collections served as a disciplinary mechanism to reinforce the nobility of the sovereign and (most often) his role as the repository of the national culture. Access was limited to an aristocratic and scientific elite. However, according to Bennett (1995), this notion of the display for the select few shifted in the nineteenth century. Instead, museums took on a vital new role—that of normalization.

The Crystal Palace or the Great Exhibition of 1851 in England is an important historical signpost in this shift. A Victorian expression of imperial might and industrial sophistication, the Great Exhibition offered an assemblage of the world in one place, juxtaposing nations in terms of art, architecture, nature, and industry. As Timothy Mitchell notes, the Great Exhibition's suggested purpose was to chart the "development of mankind" (1988, 6). In this way, the British used the exhibit to position themselves at the pinnacle of civilization, judging that which had gone before as primitive or lesser. According to McClintock, this was an opportunity to understand, rank, and consume the world's cultures and natures:

> As a monument to industrial progress, the Great Exhibition
> embodied the hope that all the world's cultures could be gathered
> under one roof—the global progress of history represented as the
> commodity progress of the family of Man. At the same time, the
> Exhibition heralded a new mode of marketing history: the mass
> consumption of time as a commodity spectacle. Walking about the
> Exhibition, the spectator (admitted to the museum of modernity
> through the payment of cash) consumed history as a commodity.
> The dioramas and panoramas (popular, naturalistic replicas of
> scenes from empire and natural history) offered the illusion of
> marshalling all the globe's cultures into a single, visual pedigree of
> world time. (1995, 57–58)

Working as visual documentation of the might of British ingenuity and mastery, the museum began to function as a new kind of space: a sphere where the working class might be civilized and made into budding bourgeois subjects (Bennett 1995). Indeed, what the museum does by Bennett's account is produce the ideal bourgeois self, which is then disseminated across the social body. Thus, the Great Exhibition marks the sea change in the role of museums: they become an "instrument of public instruction" through the mass consumption of art, culture, history, and nature (95). And it is through this consumption that people were meant to become better kinds of citizens, "to become, in seeing themselves from the side of power, both the subjects and the objects of knowledge, knowing power and what power knows, and knowing themselves as (ideally) known by power, interiorizing its gaze as a principle of self-surveillance and, hence, self-regulation" (63). Accordingly, the museum adopted the province of educating the masses (of a particular kind), through the production and display of knowledge about various disciplines, to become better citizens of the nation. At the same time as working to write on the bodies of white, Western, bourgeois subjects, it also fixed others—namely, racialized and gendered bodies—in both space and time. In other words, it became a space of governmentality.

In recent years, scholars have generated important studies that examine the politics and practices of modern museum exhibitions, taking up these important insights from the Great Exhibition and thinking through how they continue to function (or not) in the present day (see for example Bennett 1995, 2005; Glaser and Zenetou 1994; Hetherington 1997; Luke 2002a; Macdonald 1995, 2000, 2002; Macdonald and Fyfe 1996; Stine 2002). They have shown that museums work not only to display that which is known, but also to produce knowledge about the world, framing how we come to understand art, culture, empire, and nature. Producing, employing, and perpetuating "cultural master narratives" (Sherman and Rogoff 1994, xi), the museum works to circumscribe how we may encounter and understand the world. In this way, museums have played a vital role in defining linear notions of history—a progression from "primitive" to advanced, from distant to near, from history to present (Bennett 2005). As Foucault reminds us, the museum (and all those institutions that function similarly) is a memorial to our understanding of what it is to be modern: "[T]he idea of accumulating everything, of establishing a sort of general archive, the will to enclose in one place all times, all epochs, all forms, all tastes, the idea of constituting a place of all times that is itself outside of time and inaccessible to its ravages, the project of organising in this a sort of perpetual and indefinite accumulation of time in an immobile place, this whole idea belongs to our modernity" (1998, 182). In performing this modern sensibility, museums are key sites in the production and citation of notions of the nation, citizenship, empire, race, class, gender, sexuality, and, of course, nature, which function to tell the stories people know about their own and other people's past.

To this end, the modern public museum functions to tell its visitors the "truth" about the world. Museums work to produce and reiterate the stories of modernity: that history operates through an ordered and linear progress, that science will reveal the laws of nature, and that the world is knowable and made so through its representation (Luke 2002b; Pearce 1992). The museum, then, works by rendering the world visible: materializing the narratives of modernity through its collection and exhibits. This is made possible, according to Bennett (1995), by the shift from rarity to representativeness in museum collections, which allows for the world to be made as knowable to the general population. Shelved are the curiosities of the past, and in their place are those things that can be generalizable, which speak to the universal properties of the human and nonhuman world. Thus, the museum operates as a "social code" (233), where the authoritative truth about art, history, culture, or nature can be made, apprehended, and acted upon. And the iteration of these narratives depends upon the display of objects, much as it did in both the Renaissance and classical *epistemes,* which become enrolled in the service of defining the "truth" of the world through representation (Bennett 2005).

Less attention has been paid to the conflation of nature with particular cultures in natural history museums. Natural history museums might have been agents in querying the boundaries between human and nonhuman life, perhaps organizing display around the interaction between humans and nonhumans in various ecosystems. However, until quite recently, people and plants at such museums have remained quite distinct, funneled into separate halls of culture or nature. Simultaneously, an undergirding narrative of progress is ever present. By asserting that particular people and their cultures are relevant to the display of natural history, the anthropological work of these museums subtly arranges some people as closer to nature and leaves other cultures out completely. So, at the Field Museum of Natural History in Chicago, the permanent culture halls include exhibits that address Africa, Asia, and the Indigenous inhabitants of the Americas. The halls of the American Museum are similarly organized. It seems that some cultures are meant for display, while others find their artifacts in museums of industry, commerce, or art. Natural history museums have collected "exotic" cultures as they collected "exotic" nature, but only those that can be placed on the side of history rather than the present. There is a clear distinction, as Clifford (1995) has remarked, between the collector and the collected. So, there is a spatial separation of nature and culture at the same time as a temporal association between particular cultures and natures. In this vein, a natural history museum depicting Indigenous culture becomes more than a collection of art, utensils, and clothing. Instead, these objects recount a larger story about history, modernity, and nature—a lesson in cultures left behind that informs our modern understanding of progress and civilization. Often the display of these objects can render Indigenous people as anterior to modernity: as captive to a culture of the past. Thus, these objects provide the means through which the visitor can comprehend how the world functions and her place within it. It is through this representation that we come to know ourselves (Sherman and Rogoff 1994, xii).

Natural History Museums in the United States

While cabinets of curiosities and natural history museums had a long lineage in Europe, they only rose to ascendance in the United States in the 1800s.[3] Through the course of the nineteenth century, interest in natural history grew, reaching its apogee in the Civil War years and beyond, with the birth of Harvard's Museum of Comparative Zoology in 1859, the American Museum of Natural History in 1869, the Smithsonian Institution in Washington, D.C., in 1881, and the Field Museum in 1893. As Conn (1998) relates in *Museums and American Intellectual Life, 1876–1926,* each of these museums followed the Victorian practice of display to generate an understanding of nature as ordered, with the knowledge that their curators engaged in a benevolent and civilizing mission to enlighten the world about said order. And so, curators of these natural history museums wished to produce two truth effects simultaneously: first, that visitors be humbled by the majesty and complexity of nature; and second, that visitors must also understand that nature's Gordian knot could only be untangled through the labor of science (Wali 1999).

Until the early 1900s, natural history museums in the United States were sites of intellectual ferment and scientific investigation. Much like their European counterparts, museums such as the Smithsonian and the American Museum were centerpieces in the quest for knowledge around botany, biology, natural history, and anthropology. However, this role began to shift with the ascendance and proliferation of universities (Conn 1998).[4] Conn asserts that the focus on morphology that natural history museums espoused lost favor as science became increasingly fixed on evolutionary processes and later, genetic sequencing. In line with the transition to Foucault's modern *episteme,* the study of biology concerned itself with invisibility: "to know why it is that things came to look as they do," rather than being satisfied with what was visible (Hooper-Greenhill 2001, 17). This meant that the sciences of biology, zoology, and paleontology were more properly attuned to the laboratory than the curatorial shelf.

This is not to suggest, however, that natural history museums somehow became obsolete. In fact, they rolled with the punches, and continue to, rather than fading into obscurity. Thus, the 1920s ushered in a new function of museums: entertainment as well as public education. At the forefront of this change was the American Museum, with their groundbreaking construction of dioramas to depict the habitats of the world. The supremacy of the American Museum in this kind of display is still without question. At the same time, museums became the haunt of school groups, and to this day, a visit to a natural history museum seems de rigueur in most curricula. In later years, most natural history museums changed with scientific consensus, moving from a phenetic taxonomy based on morphology to a cladistic approach rooted in evolutionary relationships (Wali 1999), emphasized in the "Spectrum of Life" exhibit, discussed below. Indeed, many museums have restructured their exhibits to pay attention to advances in ecology, demonstrating the functioning of ecosystems. And more recently, natural history museums have reimagined their role again, seeking to compete with films as well as sites like Disney's Animal Kingdom, which make nature more "exciting." So, while some museums have perhaps lost their position on the cutting edge of scientific research, they still maintain potency in teaching about both nature and society. None is perhaps so important (at least in the United States) as the American Museum of Natural History.

The American Museum of Natural History

World-renowned for the production of knowledge about nature, the American Museum hosts a massive collection of 30 million specimens that attest to its importance in the game of defining nature. An authoritative source on what is defined as real, valuable, and worthy of memorialization, this museum functions as the arbiter of what counts and what doesn't in the construction of nature. As Luke asserts, "Being included in the collections of the American Museum of Natural History virtually canonizes anything as part of 'nature' and/or 'history'" (2002a, 110).

The main entrance to the museum heralds its mission: "Truth, Knowledge, Vision" is writ large above the American Museum's doors. The American Museum was established in 1869 by Albert Bickmore Smith with the support of New York's elite, including philanthropist William E. Dodge Jr., Theodore Roosevelt Sr. (father of the eventual president), and noted financier J. Pierpont Morgan. While the American Museum is now swathed in the cleansing mantle of science, its beginnings were somewhat less auspicious. Indeed, Conn (1998) has argued that rather than following the impetus of scientific inquiry, the museum's early forays in collecting demonstrated the interests of the corporate benefactors instead of the desire for a well-balanced collection. Kadanoff (2002) suggests, then, that the alliance of people like Marshall Field, the owner of the first department store, and the Field Museum in Chicago named in his honor, are not so unlikely as they might appear at first blush. Natural history museums in the United States were born of wealth, consumption, and spectacular display.

The American Museum soon got down to the serious business of assembling and housing an enormous collection of the globe's nature, which has led some to charge the museum with "institutional imperialism" (Teslow 1995, 14–15). The ascendancy of Morris K. Jesup to President of the American Museum in 1881 marked the beginning of "a golden age of exploration" between 1880 and 1930, when many of the museum's 30 million specimens were collected through a series of scientific and quasi-scientific expeditions that served to flesh out the character of the United States and the world.[5] In a book copublished by the American Museum entitled *The American Museum of Natural History: 125 Years of Expedition and Discovery,* Rexer and Klein (1995) narrate the history of these expeditions as noble and heroic endeavors on the part of human and nonhuman nature; the work of intrepid scientists who catalogue, classify, and make ordered sense of a chaotic world. But, in fact, many of the sojourns that defined the so-called "golden age" were not conducted by scientists but rather wealthy collectors who fancied themselves amateur naturalists. Further, Luke asserts that "many of its collections were gathered in wide-ranging freebooting *Indiana Jones*–style expeditions that the city's media have celebrated for decades" (2002a, 101). So, in some ways, the scientific imperative was linked to the goals of benefactors—perhaps not so different from how it is today.

The American Museum positions itself as the repository of all natural knowledge, while eliding the ways in which this process of "discovery" defined and circumscribed various cultures, practices, objects, and subjects: "Expeditions are the embodiment of this Museum's mission of discovery and understanding, throughout its illustrious past and in its continuing role at the forefront of scientific research. More than the sum of its twenty-three buildings, forty exhibition halls, and 30 million specimens, this is a place born of dreams and built on a solid foundation of intellectual inquiry" (Rexer and Klein 1995, 17). In this way, the museum can imagine itself (and hence be imagined) much as Pratt describes the identification of early naturalists such as Linnaeus or von Humboldt: engaged not in a project of imperialism but rather an ordering of the world in a "global classificatory project" where "the observing and cataloguing of nature itself became narratable" (1992, 27–28). As Rexer and Klein note, the American Museum's task is nothing less than "mapping the origin and progression of life on this planet and exploring the richness and variety of world cultures" (1995, 15). A tall order, and one that works to erase both the origins of collecting at the museum and the (neo)imperial drive to harvest the world's natures and culture and bring them back to the metropole.

Roy Chapman Andrews, Margaret Mead, Carl Akeley: each of these luminaries was central to elevating the reputation of the museum and recuperating it from its less-than-fully scientific beginnings. Roy Chapman Andrews's paleontological expeditions to Mongolia in the 1920s retrieved some of the dinosaur specimens that are still on display today; Margaret Mead, curator at the American Museum between 1926 and 1978, along with Franz Boas, her mentor and organizer of the historic Jesup expedition, are credited with the founding of modern anthropology (AMNH n.d.1). The advances in taxidermy pioneered by Akeley, culminating in the Hall of African Mammals, have made the museum synonymous with the display of natural history. Taken together, each one of these "explorers" added to the American Museum's fame and helped position it as an unquestioned space for the generation and dissemination of objective, scientific knowledge.

And the American Museum continues in this vein. Unlike Conn's assertion that research fled the museums for universities, the American Museum has maintained its scientific credentials. It employs experts in paleontology, invertebrate and vertebrate zoology, anthropology, and astronomy. The museum has been on the cutting edge of the cladistic display of natural history (AMNH n.d.2). Indeed, it can act like a university, granting PhDs and MPhils through the Richard Gilder Graduate School of the American Museum of Natural History, in association with Yale, Cornell, New York University, Columbia, and the City University of New York (ibid.). Given this evidence, I argue that not only does the American Museum circulate research produced elsewhere, but it is actively involved in the production of scientific knowledge, whether through the research projects of their own staff, partnerships with other organizations and universities, or by granting access to their extensive collections to a host of scholars.

In recent years, the display of scientific prowess and the bounties of nature have given way to what some have termed "edutainment." The American Museum does not stand outside of this trend. Indeed, the Hall of Biodiversity, as well as other recent permanent exhibitions, like the Spitzer Hall of Human Origins and the Rose Center for Earth and Space, reflects this new turn in exhibitionary practice quite well. Each of these exhibits employs multiple forms of media, computer technology, interactive displays, color, and hands-on items to tempt museumgoers. These permanent halls are complemented by the various films and traveling exhibitions that use these technologies and keep visitors coming back again and again to see what is new. In an ironic turn, as museums seek to make themselves more current and palatable, adopting the tools of theme park culture to attract more and more patrons, Disney mimics museum practice at the Animal Kingdom in DinoLand U.S.A. (for a fuller explanation, see chapter 2). Some celebrate this change (M. King 1991), while others lament the passing of what is seen as a purely scientific approach to display (Kadanoff 2002). In either case, it seems as though the dusty display case is becoming a thing of the past. And so, the American Museum is an uneasy mix of old and new, dioramas and GIS maps, dinosaur bones and IMAX movies.

Perhaps it is their prestige as a learned institution combined with their savvy in recognizing and seizing new trends of display that has helped maintain the American Museum's iconic status. From its beginnings on the outskirts of Manhattan, the American Museum has since become a huge organization of 1,250 employees and departments ranging from Exhibitions and Multimedia to Invertebrate Paleontology, from Physical Anthropology to Astrophysics. The American Museum has emerged as not only a key producer of knowledge about nature, but also a foundational site for its consumption. According to the staff I interviewed, this museum stands as "the most visited museum in the country," second only to the Disney theme parks as the largest tourist attraction in the United States (anonymous interview 1, 2006). The museum receives approximately 2.5 to 3 million visitors per year, half from the New York Metropolitan area, 30 percent from other places in the United States, and 15 percent from other countries. Approximately 15 percent are children on school trips (interview 1). With total assets of $1,051,342,672 in 2009 (Grant Thornton 2009), the

American Museum is a global mover and shaker in the definition and display of nature; it is, as it proclaims in a recent fundraising e-mail, "where science meets society."

The Hall of Biodiversity

The Hall of Biodiversity opened in 1998. The mounting of this permanent exhibit was funded through the National Science Foundation, NASA, the City of New York, and various foundations, as well as more unlikely bedfellows like Monsanto, Mobil, and Bristol-Myers Squibb (discussed further below). Curators and staff defined the exhibit as the first "issues hall" the museum has ever mounted.[6] Housed in an 11,000-square-foot space, this permanent exhibit was intended, according to its curators, to "put a public face on conservation" (anonymous interview 1, 2006). In a dramatic turn in the epistemological function of museums, this hall, along with similar exhibits at the Field Museum and the Smithsonian Institution, seeks to

Map of the Hall of Biodiversity at the American Museum of Natural History in New York City.

present information about processes as well as the display of objects, and in so doing comments upon the state of human and nonhuman relationships (Wali 1999).

The exhibit is structured around four questions, each of which corresponds to a part of the hall. "What is biodiversity?" is explored in the "Spectrum of Life," the wall dedicated to the cladistic tracing of evolutionary pathways. "Why is it important?" is examined in the next section of the exhibit, the Dzanga-Sangha diorama, which depicts a rainforest in the Central African Republic. The next question, "how is it threatened?" is addressed in the area entitled "Transformation of the Biosphere," which considers the causes and impacts of human interaction with the environment. Finally, the exhibit asks "what can we do?" in the part of the hall called "Solutions," which contains a series of wall plaques and computer monitors that demonstrate the actions that ordinary people can take to stem the tide of environmental disaster. Although each part of the exhibit is meant to answer its respective question, they all highlight the interconnection and plenitude of life, which is one of the key goals of the Hall of Biodiversity. The curators of this exhibit utilized technology to make the Hall of Biodiversity accessible, interesting, and powerful, in line with the next generation of museum display. As one of the lead curators noted in an interview with me, "exhibitions themselves, in their entirety, are works of art, at least the good ones are" (anonymous interview 2, 2006). As such, the Hall of Biodiversity combines models, specimens, a twenty-first century diorama, films, computer stations, simulations, and remote sensing/GIS visualizations to give its visitors a sense of biodiversity and its crisis.

Ralph Appelbaum Associates was the interpretative museum design firm hired to stage the hall, for which they won an environmental graphic design award. They assert that the exhibit is organized to elicit two effects: awe in the face of nature and a clarion call to do something about its destruction (Society for Environmental Graphic Design 1999). In terms of the first effect, the visitor is confronted with the notion of *nature as church*—a place for spiritual rejuvenation, transformation, and reconnection. Of course, this sentiment is not at odds with much environmentalist thought, which harks back to the romantic period from which some strains of modern U.S. environmentalism emerged. Indeed, in his recent book on religion and nature entitled *A Greener Faith,* Gottlieb notes that the spiritual impulse compels the need for action: "It is precisely because for many of us the encounter with the rest of life on earth—indeed at times with the universe as a whole—provokes profound feelings of awe and reverence, mystery and serenity, that we commit ourselves to its protection" (2006, 151). This is evidenced in the Hall of Biodiversity, where the exhibit is dimly lit, evoking a sense of quiet respect in the viewing of nature. It is peppered with quotes from the luminaries of environmentalism that testify to the importance of a personal relationship with nature. Thus, Henry David Thoreau is here, announcing, "We need the tonic of wilderness." Robert Louis Stevenson is quoted as saying "that the forest makes a claim upon men's hearts, as for that subtle something, that quality of the air, that emanates from the old trees, that so wonderfully changes and renews a weary spirit." Rachel Carson, Baba Dioum, Oren Lyons, Paul Ehrlich, and Theodore Roosevelt are also present, advising us to stand in awe of nature (and save it) before it's too late. Further, the hundred-foot-long "Spectrum of Life" as well as the rainforest diorama reproduces nature on such a scale that it is difficult to be anything but humbled before it.

The second intended effect—encouraging action—is addressed through the parts of the exhibit that show how biodiversity is threatened and offer examples of what we can do about it. Thus, the rainforest diorama, the wall entitled "Transformation of the Biosphere," and the "Solutions" section all speak to the current crisis in biotic life and provide the visitor with manageable steps to ameliorate the loss of nature. In this way, the exhibit achieves its intended aims.

A range of visitors have consumed these aims. I tried on numerous occasions to access both statistics on, and surveys of, the visitors to the Hall of Biodiversity to no avail. As such, I can only rely on my observations of the patrons present in my visits to the museum, though I suspect the people who frequented the hall during my research were not unusual and might be thought of as representative of the average museum visitor. In my fieldwork at the museum, the most prevalent visitors were children: school groups primarily, but also what appeared to be nannies and parents with their children. For the most part, these children were young, somewhere between ten and fourteen, though toddlers and older teenagers were also present. The school groups entered the hall like a cacophony of birds, raising the decibel level considerably during their stay. There were also young people who seemed to be engaged in some sort of research or art project, sketching elements of the "Spectrum of Life." In addition to these visitors, there were smatterings of adults who wandered through the hall stopping to look at various items of interest. On the whole, they were an ethnically diverse crowd, and many different languages were spoken, although in hushed and reverential tones (at least by the adults).

Introductory Film: Life in the Balance

The introductory video to the Hall of Biodiversity, entitled *Life in the Balance,* is strangely situated; indeed, one might only encounter the film at the end of the exhibit, particularly if the visitor follows the structure laid out by the questions it asks.[7] In fact, very few people seem to watch the movie, with the exception of a mother who insisted that her child stay until the end. Tom Brokaw, trusted newscaster and voice of authority, narrates this ten-minute documentary film accompanied by images of magnificent and majestic nature interspersed with nature as degraded and damaged, seemingly meant to inspire both awe and guilt. This film serves to underline the main thrust of the exhibit, exploring in brief what biodiversity is, how it is threatened, and what can be done about the shared peril we (human and nonhuman life) face.

The film begins by defining biodiversity as the richness of life, indicating that it is the "abundance, variety, and interconnection" of biodiversity that characterizes our world. Humans are included as part of this biodiversity, and the film stresses our reliance on nature. For example, the film explains that people need biotic diversity for food and water, traditional and modern medicine, and shelter. We must also have nature to satisfy the affective need to "retreat to the wild places to restore ourselves." Brokaw tells us that life is everywhere on the planet, "but life is everywhere under threat." These hazards, the viewer is told, stem from habitat loss, pollution, overpopulation, and overconsumption, represented through images of clear-cuts, urban sprawl, and industrial wastelands. Extinction is identified as at an all time high, and the "pristine" habitats of the world (with special focus on the equatorial rainforests as the seat of biodiversity) are disappearing. Life, as the story goes, is truly in the balance. But the film also works to inspire hope. The viewer is told that "the solutions are clear": stabilize population growth, reduce consumption and waste, set limits on fishing and forestry, and develop clean technologies. If these policies are implemented, we will see biodiversity, now under extreme threat, brought back into harmony (an equilibrium, one would presume, subject to scientific definition). As the film ends we are told that "the future of life is in our hands"; a caption flashes at the bottom of the screen: "Replace, Replant, Conserve."

Oscillating between scientific authority and apocalyptic narrative, this film does important discursive work to frame the rest of the exhibit. It tells the story of an unproblematic truth of the global environment, without differentiation, without specificity, and without the complicated articulations of responsibility and justice. It is also prepackaged in a form that is familiar, acting almost as a sound bite that abbreviates the story of nature, its value, its threats, and solutions. The film enacts a visual display that reinforces the dominant tropes of a world in crisis. It also positions "us" at the center of this drama, as both the purveyors of mass environmental destruction and the medium through which it can be saved. This centering of human action runs throughout this exhibit, and indeed, through all the case studies in this book.

The "Spectrum of Life": Mapping Populations in Life and Death

The "Spectrum of Life"—the part of the exhibit that answers the question "what is biodiversity?"—is a one-hundred-foot-long wall that quite literally maps life on earth. It is a massive cladogram that represents 28 groups of organisms and their evolutionary connection through the display of 1,500 specimens and their replicas (AMNH n.d.1). Host to giant lobsters, turtles, spider crabs, starfish, a manta ray, a kangaroo, birds, butterflies, octopi, insects, ferns, fungi, a plethora of different bacteria, and much more, the "Spectrum of Life" assaults the senses with the diversity of life on earth. One reads the cladogram from the top down and from left to right. At the top, one finds the majority of biological categories of life, eukaryotes and animals, which then branch off into subcategories of plants, arthropods, vertebrates, tetrapods, and amniotes. These are then further divided into categories that are more familiar to the nonscientist: sponges, mollusks, roundworms, mammals, green algae, and the like. This charting of nature's evolutionary path works to induce awe in the face of the monumental achievements of nature, and of science in classifying them so clearly. It stands as nothing less than a biopolitical map of all life. This attempt to map both the living and the dead should come as no surprise; as entomologist David Grimaldi notes of the insect collections that the AMNH, "We have half of all the described diversity on earth in one room here at the museum—in many ways, *we are biodiversity*" (qtd. in Wallace 2000, 125, my emphasis).

Like the Great Exhibition of 1851 discussed above, the "Spectrum of Life" is a monument to McClintock's (1995) notion of panoptical time.

The "Spectrum of Life" at the American Museum of Natural History, a hundred-foot cladogram depicting the evolutionary connections between life on earth.

Through this cladogram, the evolutionary process can be made visible and understandable. As McClintock tells us in reference to the charting of the "Family of Man" in the nineteenth century, "The tree offered an ancient image of a natural genealogy of power" (1995, 37). But the "Spectrum of Life" is more than simply a genealogical tree. It is a complex representation of the ways that genetic derivation has shaped the bacteria, fungi, plants, and animals we know today.[8] It takes the invisible processes of DNA and renders them visible, knowable, and, by virtue of their visibility, savable. The "Spectrum of Life" shows the evolutionary web as a total history, from the origins of life in early bacteria flowing into arthropods such as mammals and mollusks. This imaged representation of the work of science is accompanied by a flood of information at each of the computer kiosks, if the visitor is inclined to learn more about the various species represented. Documented here is the number of known species (for example, there are one hundred thousand known species of fungi and lichens), where these species live, their role in ecosystems and for human needs, their characteristics, their size, and examples of common species in each clade.

Each element and attribute of the clade is noted, recorded, and made available for consumption. Indeed, once displayed, this map of 3.5 billion years of genetic variation can be consumed as a spectacle. This compression of time into a new version of McClintock's "Family of Man" allows natural populations to be catalogued, assessed, and understood—a visual representation of the power of science that brings order to global nature. This is a map of all life, and all possibilities for life.

Of course, one of the central ways the American Museum could be described is as a repository of scientific biopolitical knowledge, a node for the collection of information about nature. Thus, the "Spectrum of Life" is but a visual representation of this labor: a means of truth-telling to herald the import of both the American Museum and the science that underpins its endeavors. The counting, measuring, classifying, ordering, and display within the museum demonstrate the desire to know about life in all its dimensions. Indeed, this enterprise has taken on more urgent proportions as biodiversity threatens to disappear before it can be catalogued and assessed. One curator of the hall notes: "We have seen that the world is in the midst of a new surge of extinction—one that is taking some 27,000 species a year from the planet, many of which are doomed before anyone has the opportunity to find them, to study them, to name them, and perhaps to realize their worth" (Eldredge 1998, 181). The impetus to catalogue the world's species, particularly those that are not yet known, gives further significance to the scientific research that generates information on the uses of these species and their worth to humans. The worth of these species—aesthetic, scientific, and economic—is in part what drives the pursuits of the American Museum. The museum works to measure, assess, catalogue, determine threats to, and provide solutions for biodiversity, serving to construct and circumscribe its problems and self-evident solutions. And this, in essence, is an expression of a particular kind of scientific green governmentality.

In terms of its collections, the provost of the American Museum makes clear that continuous scientific research is necessary if the world is to avoid the impending biological catastrophe: "It is more important than ever to know what living things the world contains and to understand their interrelations, at both the genetic and species level. . . . Otherwise, we can't make truly farsighted decisions about how we manage human impact on the environment" (Michael Novacek, qtd. in Rexer and Klein 1995, 29). In this vein, the scientific departments at the American Museum continue

to embark on expeditions to collect specimens, but they also work on existing collections to determine the character and function of species within
the global biosphere. Through their efforts, the American Museum seeks
to provide objective knowledge about the health and well-being of human
and nonhuman populations so that sound decisions can be made about
how best to manage global nature. "[G]overnments, scientists, industry,
and policymakers throughout the world must take complex collective
actions to address species loss, and these actions must be informed by
accurate scientific information about species diversity" (AMNH 1998a).
And so, the "Spectrum of Life" represents the physical manifestation of
this work, doing three things simultaneously: (1) demonstrating the diversity of life and hence the need to continue the project of classification; (2)
working as an actualization of the mastery of science and success of this
project; and (3) allowing visitors to comprehend this diversity, in all its
glory, so that they might support the cause to save it.

Moreover, part of the way the American Museum organizes its biopolitical knowledge is through the literal re/presentation of life in death.
It is through the collection of the more than 30 million specimens that the
museum houses that it makes life through death; it seeks to make knowable the secrets of nature through actually killing it. This is evidenced in
the Hall of Biodiversity through the "Spectrum of Life" as well as a glass
case nearby entitled the "Crisis Zone" discussed in the introduction to this
book. Through the display of these specimens (some 1,500 in the "Spectrum of Life" alone) visitors are called to both connect with nature and
learn about its meaning. Thus, the visitor is to accept that this killing of the
specimen was necessary: a vital part of re/presenting it so as to further
knowledge about life. As the museum notes, these specimens provide an
"invaluable record of biological communities and ecosystems over time"
(AMNH 1998b), which is then reexamined and deployed to produce new
knowledges about the environment. So, through killing "we" can avoid
extinction. Not only do these collections provide us with information about
the past, but also the future: hence, in death there is life. And it is through
this death that life can be managed more and more effectively.

"Dzanga-Sangha Rainforest Diorama"

Following the exhibition plan, the next part of the Hall of Biodiversity that
the visitor encounters is the "Dzanga-Sangha Rainforest Diorama." At ninety

feet long, twenty-six feet wide, and eighteen feet high, this representation of
the Central African Republic (CAR) rainforest seeks to replicate how this
place looks, sounds, and even smells. Eerily similar in some ways to Disney's
Tree of Life in the Animal Kingdom (though, of course, it aims for facsimile rather than fiction), this diorama includes some five hundred thousand leaves that were handmade to flesh out the replica of the rainforest
(AMNH n.d.1). Moreover, it displays over 160 different species of plants
and animals (though most of the animals are virtually impossible to see).

The diorama is based on an actual place, the Dzanga-Sangha Reserve
in the CAR, which was established in 1988 and legislated as a national buf-
fer zone for the Dzanga Ndoki National Park in 1991. Up to its establishment, the economy of this area rested upon forestry, and the population
of the closest town, Bayanga, followed the cycles of the lumber industry.
With the advent of the reserve, the town's economies have shifted to include
trophy hunting and ecotourism as well as forestry (Remis and Hardin 2009).

As one of the co-curators remarks, "All rain forests are different, but the
Dzanga-Sangha rainforest, with its spectacular diversity, is special" (Joel
Cracraft, qtd. in Lyman 1998, 9). It is this singularity that answers the
question that frames this part of the Hall of Biodiversity: why is it important? The Dzanga-Sangha rainforest and other biodiversity hotspots are
important because, in their remoteness and rarity, they offer a wealth of
biodiversity. The intent of the rainforest diorama, according to one of the
curators of the hall, is to invoke a sense of mourning for the loss of this special habitat by depicting nature in three states: "Pristine, altered by natural forces, and degraded by human intervention" (anonymous interview 2,
2006). The lesson here is that nature is under threat from human activities: international forestry to be sure, but also poaching, traditional hunting, illegal diamond mining, and the clearing of land for settlement.

The visitor is asked to engage viscerally with the exhibit. Rather than
the more conventional dioramas for which the American Museum is so
famed, the Dzanga-Sangha rainforest is represented without a protective
glass shield. Visitors are "invited behind the glass wall that usually forms a
barrier between viewers and the scene depicted, and thus become a part of
this ever-changing and severely threatened ecosystem" (AMNH 1998b).
In this way, the visitor is asked to experience the rainforest—the representation of it becomes its reality, much like many of the sites encountered
at Disney's Animal Kingdom. But it also serves to ensconce the viewer in
the drama of the rainforest's destruction, gazing upon the stages of its

degradation. The impetus for action here is clear: "The goal of the Museum remains the same—to bring to visitors the beauty and purpose of the natural world in the hope that we will all be encouraged to understand, respect, and preserve it" (Lyman 1998, 127).

As mentioned above in the history of the American Museum, this institution has become synonymous with display through dioramas. Named as "art to conceal art" (Quinn 2006), dioramas at the American Museum have long functioned to tell the "truth" of the various habitats of the world. Indeed, the production of dioramas has always been linked to notions of "authenticity": they stand as a truthful representation of the world to be gazed upon, and through that gaze, nature becomes understandable and consumable. Haraway dissects the role of the diorama superbly in her work on Akeley and the Hall of African Mammals: "A diorama is eminently a story, a part of natural history. The story is told in the pages of nature, read by the naked eye. The animals in the habitat groups are captured in a photographer's and sculptor's vision. They are actors in a morality play on the stage of nature, and the eye is the critical organ" (1989, 29). Akeley, one of the pioneers in the development of dioramas for the American Museum, noted the importance of the creation of photographic replicas of the places you depict: "Otherwise, the exhibit is a lie and it would be nothing short of a crime to place it in one of the leading educational institutions of the country" (Akeley, qtd. in Haraway 1989, 40). Indeed, it is through notions of the authentic that we come to know the "truth" of our current condition: we find the vehicle to mourn what we have lost (Urry 2002). The Dzanga-Sangha diorama continues with this tradition; Brian Morrissey with the Exhibition Department of the American Museum noted the expedition to the Dzanga-Sangha "walked in the forest in the shadow of Akeley" (1998, 56). And so, even though the visitor stands in a museum in New York instead of a rainforest in Central Africa, the diorama attempts to make a connection with nature, to turn on an affective need to encounter the wild. These elements make dioramas a potent vehicle for telling the "truth" about nature.

Of course, many of the specimens displayed in the dioramas at the American Museum were gathered through the institution's various expeditions. Although there has been a move away from the production of dioramas, both because of their static nature and because the specimens they require are often endangered (Quinn 2006), museum staff speak fondly of the scientific expeditions that generated these dioramas as a necessary

respite. Invoking travel narratives, and sometimes the language of imperial adventure, the expedition is characterized as a rejuvenation in the
wilds of nature: "Perhaps it is because the world had become a smaller and
more crowded place, and these scientists, unlike those who spend their
careers entirely in the laboratory, are in search of boundlessness and soli-
tude—commodities almost as precious as any new species they might discover. 'Out here,' Michael Novacek, the museum's provost, has said about
the Gobi, 'you are away from telephones and fax machines. You have the
time just to think'" (Rexer and Klein 1995, 14). Indeed, the expedition is
framed as a kind of psychological wellspring, an exploration of the mind as
well as of nature (14).

This kind of experience is evidenced in the book *Inside the Dzanga-
Sangha Rain Forest* (Lyman 1998), a children's book that charts the mounting of this aspect of the Hall of Biodiversity. It is here that we find Pratt's
"contact zone" (1992, 6): a place of the mingling of distant cultures, with
the encounter often resulting in relations of domination, but also transculturation. *Inside the Dzanga-Sangha Rain Forest* deploys the narrative of
the intrepid scientist who gives over his or her health and well-being to the
pursuit of truth. Thus, scientists endure primitive accommodations, fate-
ful encounters with elephants, a nightly barrage of flying insects, and "mystery meat" prepared by the locals (Lyman 1998, 30, 54, 91, 107). Even
though these trials are meant to elicit horror and delight, the reader is re-
assured that they were necessary to the production of truth about the rainforest. As the scientists assert: "The future of Dzanga-Sangha, like the
future of rain forest areas around the world, depends on people's understanding, appreciating, and learning to conserve their natural wealth and
wonder. We set out on our expedition to help do all these things" (20).
And it is a seductive story, weaving a travel narrative with pictures and text
on the animals, plants, and insects they encountered and collected. But it
is also a story of the importance of science and its ability to define the
natures and cultures of distant lands. As Luke remarks, "Empire is naturalized as nature is, at the same time, imperialized" (2002b, 131).

However, the story of the Dzanga-Sangha Reserve that the AMNH
represents is not simply one that can be dismissed as yet another obvious
example of the imperial endeavor. The site itself is an example of an integrated conservation and development project (ICDP), an approach that
sought to move away from "fortress conservation" that was characterized
by the displacement and policing of local populations (Adams and Hulme

2001; Hardin 2000). Since its beginning the reserve has been managed through the Dzanga-Sangha Project, a partnership between World Wildlife Fund for Nature and the government of the Central African Republic. It has also received funding from both the German and U.S. governments as well as the World Bank (Blom 2000). In line with the goals of ICDP, the project operates the reserve as a space of multiple uses: sustainable forestry, hunting and gathering, safari hunting, ecotourism, and scientific research. The exhibit celebrates this approach. In the text and video accompanying the diorama, the museum asserts that conservation must take place at the behest of, and be managed by, local peoples. As the plaque on "integrated management" that is part of the Dzanga-Sangha diorama attests, "By giving people a stake in the management and benefits that arise from protected areas, many government and conservation organizations are finding increased success in preserving biodiversity." This is certainly an approach that is preferable to the heavy-handed strategies of other conservation efforts that relied on removal rather than power sharing. But how successful is the Dzanga-Sangha Reserve in both preserving biodiversity and offering opportunities for both local participation and income generation?

The results appear to be mixed. The Central African Republic is one of the poorest countries in the world, and formal employment is often scarce. The promise of jobs as guides, trackers, or guards has brought an influx of people to the Bayanga area, increasing the population exponentially, from a thousand people to ten thousand in 2004 (Remis and Hardin 2009). However, the jobs provided by the project are much fewer: fifty full-time positions and upwards of seventy-five part-time positions (Noss 1997). Thus, logging remains the largest employer in the area, and the reserve project, according to Hardin, "has not been viewed by the growing local community as successful in providing adequate alternatives to logging income or bushmeat" (2000, 56). Indeed, there is some suggestion that the very nature of the mixed-use vision espoused through the ICDP exacerbates both environmental destruction and social inequality:

> The confluence of logging and conservation activities may foster concentrations and habituation of elephants and gorillas that render them more vulnerable to volatile poaching dynamics. Moreover, social conflict has cropped up when conservation programs have had difficulties in achieving multiple human needs, poverty alleviation, and biodiversity conservation goals. Some of

these tensions can be seen in the RDS project's repeated failure to
give credit to local communities for their conservation successes.
(Remis and Hardin 2009, 1593)

Noss links this failure to a lack of land tenure in the CAR, where all land is
owned by the state. He suggests that "[s]ince colonial times local people
have been ignored as governments award management and exploitation
rights for vast areas to external organizations: concession companies in
the 1890s, commercial societies in the 1920s, logging companies since the
1970s, safari hunters in the 1980s and conservation organizations in the
1990s" (1997, 183). For Noss, then, the ICDP is yet another iteration of a
legacy where local people are excluded from decisions about their land.

But it is also a space of encounter, precisely because it has not followed
the model of conservation where people must be evicted to save biodiversity. As such, it might offer an interesting reconceptualization of the relationships between humans and nonhumans, denying the separation of
nature and culture that is so present not only in the protection of biodiversity, but throughout the many sites and spaces described in this book.
As Remis and Hardin contend, the Dzanga-Sangha Reserve is "a site where
humans and wildlife effectively maneuver around each other, creating new
interspecies and intercultural relations of forest use. Such precarious accomplishments of transvaluation should not go uncelebrated when they succeed
or unlamented when they fail. They are beacons from forest communities
whose interactions with industry and conservation reveal cosmopolitan
cultural ecologies that enable varied and dense human interactions with
wild animals no longer found elsewhere" (2009, 1594). And so, I don't
think we can simply castigate this ICDP as the latest version of the colonial contest. Rather, it is a much more complex place, that both offers the
potential for new definitions of conservation at the same time as it seems
to fail on its goals for both biodiversity protection and local development.
And to stay with the complexity here is, I think, the important part. Because integrated management is not an innocent endeavor. And to some
degree, the exhibit signals this complexity. It notes that the greatest threat
to this ecosystem is Western demand for consumer products like dia-
monds and wood, and it speaks to conservation problems like increased
population, subsistence agriculture, and poaching. It also notes that conservation can itself be a problem: when elephant populations increase
because of such efforts, they trample the forest. But what the diorama pays

less attention to is the result of the ICDP for local populations. It doesn't address whether their lives have improved through the establishment of the reserve, how they connect to the process of decision-making, or what their view of the land they inhabit might be. Rather, while the exhibit names their importance in the process of sustainable development, the actual peoples of this parcel of the Central African Republic seem to fall away, the materialities of their lives less important than the role they fill as green ambassadors to the ICDP.

"Transforming the Biosphere"

The notion that the rainforest diorama puts forward—namely that people are both the cause of and solution to environmental ills—is explored more fully in the next part of the exhibit, entitled "Transformation of the Biosphere." This wall, which runs the width of the exhibit, is made up of text, images, and videos that aim to answer the question "how is biodiversity under threat?" This section also contains a discussion of the "Sixth Extinction": the prediction of extinction on a scale equivalent to the annihilation of the dinosaurs, and through which the world will experience the loss of half its species in the next thirty years (AMNH 1998a). As such, there are numerous panels that show various human activities that have degraded the environment: population growth, agriculture, urbanization, overfishing, and deforestation. The effects of these actions as detailed in "Transformation of the Biosphere" are habitat change and disease, the introduction and proliferation of alien species, coastal pollution, cultural extinction, biotic extinction, and importantly, loss of the services that healthy ecosystems provide. The effect of this wall is quite dramatic. Its images of environmental apocalypse combined with subtitled silent films reinforce the already hushed surround, and generate a feeling of dread in the face of all this destruction.

But rather than immersing us into a primeval rainforest to tell the truth of nature under threat, this section of the exhibit names these problems through the language of science. One of the ways through which the knowledges produced by the natural sciences become unquestionable is through their claim to expertise and authority. It is the distance from which scientists speak that lends them their legitimacy. In this way, science has become important to the construction of modalities of power: "[T]he grounding of authority in a claim to scientificity and objectivity establishes in a unique way the distance between systems of self-regulation and the formal organs of political power that is necessary within liberal democratic rationalities of government" (Rose 1996, 156). Thus, as an American Museum curator has indicated, "[a]ll the easy questions have been answered. The only ones left are difficult" (qtd. in Rexer and Klein 1995, 14). This statement necessitates an authoritative scientist to define the "truth" of nature for the rest of us. As Paul Rutherford (1999, 37) has asserted, science—and more specifically ecology—has become both the means of legitimation and the object of particular forms of governmentality.

In "Transformation of the Biosphere" we encounter undisputable facts as told by science. The importance of science is most evident in the discussion of the "Sixth Extinction" and ecosystem services. In an elegiac turn, this wall displays all the species known to have become extinct since the 1500s on a black wall reminiscent of the Vietnam Memorial in Washington, D.C. It is almost too many to count. However, this is but a drop in the bucket if we continue on the path to the "Sixth Extinction." A plaque in this section of the hall outlines the gravity of this situation:

> Not since the end of the age of dinosaurs—the great extinction of the Mesozoic Era—have species disappeared at such a rapid rate. As ecosystems recover after episodes of mass extinction, evolution eventually produces new species. But evolutionary recovery can only begin when the cause of mass extinction itself goes away, and the recovery process takes millions of years. Not until humanity realizes the effect that our massive transformation of the biosphere is having on the world's species, and not until we decide to take corrective action, will there be any hope of stemming the tide of the Sixth Extinction and allowing biodiversity to recover.

What is the impact of this loss? Many are cited, but what is most emphasized is a loss of ecosystem services. Ecosystem services are the benefits that the earth provides us as humans: carbon sequestration, water purification, photosynthesis, decomposition of waste, production of food, flood control, renewing the fertility of agricultural fields, pharmaceutical products, and so forth (see, for example, Daily et al. 2000). There is also the possibility that the genetic information inscribed in biodiversity, yet to be discovered, will provide some benefit in the future. The language of ecosystem services, now discussed as a form of natural capital (ibid.), has been deployed in recent years to give a kind of economic value to the work of nature—to put it in terms that make the conservation of biodiversity a human necessity.

The exhibit uses the notion of ecosystem services to drive home the interconnectedness of all life: that we cannot exist if we continue to decimate our environments. The emphasis on ecosystem services in the hall works to produce and reinforce the notion of global threat. As Eldredge notes:

> What is far less well understood is the other side of the coin: the impact of the *global system* on us. Because we are still stuck with the notion that we have escaped the natural world, few of us see the dependence that our species truly has on the health of the global system. The main reason we should fear the Sixth Extinction, I truly believe, is that we ourselves stand a good chance of becoming one of its victims. . . . We might well avoid literal biological extinction—but our cultural diversity, and, for the developed nations, our high standards of living, are very much at risk. (1998, 151, emphasis in original)

This threat legitimates intervention on a global scale, to manage the biodiversity hotspots (all in the tropical countries of the South) to maintain biospheric equilibrium and ensure that the ecosystem services keep on giving.[9] In highlighting ecosystem services like the pharmaceutical benefits of various plants, the exhibit functions in part to commodify nature and authorize its own role as an expert on the functional aspects of biodiversity. Moreover, as Luke contends, "The expansive reach of transnational commerce gives scientific agencies and national governments all of life's biodiversity to administer as 'endangered populations'" (2002b, 133).

When one looks at the sponsors of the Hall of Biodiversity, it is not difficult to see the connections to transnational corporations and the ways these sponsors may have shaped the information that is presented. Among the "founders" of the exhibit—meaning its most generous donors—is the agricultural biotechnology giant Monsanto. That Monsanto might be named as a key agent in biopiracy, as a purveyor of the kind of herbicides that work to lessen rather than enhance biodiversity, and as a modifier of the genetic basis of life for profit, does not register in this exhibit. Indeed, Monsanto's "terminator seeds" have been the topic of much scholarly and activist work, which considers not only its role in the production of monocultures—the antithesis of biodiversity—but also the ways that this corporation works

to erode land tenure and agricultural practices of the world's farmers, particularly in the Global South (see, for example, much of Vandana Shiva's
body of scholarship). These efforts have led some to name Monsanto's
agricultural innovations "accumulation by dispossession" (Prudham 2007).
The fact that these technologies are framed within a discourse of poverty
reduction and hunger alleviation (Glover 2010)—as mechanisms to lift the
world's poor out of starvation—makes them seem all the more pernicious.

But Monsanto is not the only sponsor of the Hall of Biodiversity
accused of limiting natural variation. For example, Bristol-Myers Squibb,
a pharmaceutical giant and donor to the exhibit, has patented a synthetic
version of Taxol, taken from the bark of the Pacific yew tree, as a potent
anti-cancer drug (Chacko and Sambuc 2003). The fact that Indigenous
peoples have used elements from this tree and many thousands of others
for curative purposes for generations does not generate any rights, as traditional ecological knowledge loses out to the scientific practice of harvesting biodiversity. Indeed, as Moran documents, "Of the 120 active
compounds isolated from higher plants and used today in Western medicine, 74 percent have the same therapeutic use as in native societies"
(2000, 133). But this knowledge holds little power in the circuits of international capital generated through biodiversity, infected as intellectual
property rights are with racialized notions of what counts as "real" science.
And so, corporations like Bristol-Myers Squibb and Monsanto are given
free rein to patent and commodify parts of nature. Following Rose (2006),
it is interesting to note that Monsanto and Bristol-Myers Squibb seem to
be engaging in new forms of green governmentality, through the deterri-
torialization and reterritorialization of the genetic basis for life. So, these
companies do not simply destroy but are involved in an ongoing project
to change, and they would argue improve, basic genetic structures. But the
work of these companies represents a governing of, and intrusion upon,
the biosphere, something that the Hall of Biodiversity would seem to contest. At the same time, it erodes the ability of local communities to utilize
their knowledge of the land in ways that, for thousands of years, have actually enhanced biodiversity.

Further, another funder of the exhibit is Mobil, whose parent company
is Exxon. One cannot forget the Exxon Valdez spill and its ravaging of the
coastal ecologies of Alaska. So, it begs the question: How is it possible that
these are some of the main funders of an exhibit to educate and promote
action on the biotic diversity of the globe? What of the irony? These funding

linkages are a part of the exigencies of the nonprofit world, of this there is no doubt. But why *these* corporations? And how do they shape the exhibit? It seems to me that what gets reified here is a distinction between "real" scientific research and local practice, naming Western science as a legitimate and improving source of knowledge about the natural world, whereas what is awkwardly termed "traditional ecological knowledge" is rendered a place to begin research—a knowledge network to mine—rather than a valuable end in itself. That this distinction is racialized and inscribed along circuits of colonialism is indisputable. Indeed, the exhibit celebrates this scientific pursuit, without ever having to address the thorny issue that it is these very corporations that reduce biotic diversity, lessen food security, and limit access to local nature-cultural practices. Moreover, receiving sponsorship from Bristol-Myers Squibb and Monsanto in particular points also to the ways in which the knowledge produced via the scientific work displayed in this exhibit is of central importance to those corporations that generate profits from knowing where, when, and how to access the richness of life. That this knowledge is used to dispossess local, and often Indigenous, peoples can have no name here. Instead, what is important is the biodiversity crisis, which must be mapped and its bounty harvested before it disappears. This fits nicely in both the museum's articulation of the importance of the scientific pursuit and with the biotech companies' pursuit for ever-higher profits.

Moreover, the exhibit's emphasis on ecosystem services points to its implication in the production of specific kinds of knowledges that are formed to benefit the health and well-being of the global population, couched within the language of neoliberalism. But the punch line here is that nature can be imagined as only existing to minister to human needs: a set of benefits that cannot be lost. Luke (2002b, 141), in his work on the Missouri Botanical Garden, remarks that ecosystem services, then, define what is important in nature: "As an environmental engine, the earth generates 'ecosystem services,' or those products and functions derived from natural systems that human societies perceive as valuable, faunapower and florapower reserves. This is what must survive."

If this is what we stand to lose, so the exhibit argues, where can we place responsibility for this probable calamity? At the feet of *the entire* world, both North and South. As a species, we are causing this monumental global disaster that, because of the interconnection of all life, spells apocalypse for us as well:

Now a single biological species, our species, *Homo sapiens*, is disrupting ecosystems and driving species to extinction in the new wave of mass extinction. Why not just let the Sixth Extinction run its course? After all, evolution ultimately creates new species that become the players in newly rebuilt ecosystems. The answer is simple: New species evolve, and ecosystems are reassembled, only after the cause of disruption and extinction is removed or stabilized. In other words, *Homo sapiens* will have to cease acting as the cause of the Sixth Extinction—whether through our own demise, or, preferably, through determined action, before evolutionary/ ecological recovery can begin. Our fate is inextricably linked to the fate of Earth's species and ecosystems. (Eldredge 1998, 66)

Although there is emphasis placed in this exhibit on the role of consumption in the North, the direst pronouncements about the roots of this crisis are linked to overpopulation, and, by extension, the Global South. As a plaque in the exhibit proclaims, "with the explosion of the human population following the invention of agriculture around 10,000 years ago, our demand for and impact upon biological resources dramatically intensified." Locating the problem in the notion of overpopulation is demonstrated by one of the more popular parts of this exhibit: the visualization that depicts the population increase from 1 A.D. to projected estimates at 2025 A.D. This visual offers museumgoers the opportunity to lay blame at the fecundity of the other, with some remarking, "There's too many people in the world. Look at China." While this video goes on to remap the world based on CO_2 consumption, resource use, and wealth—all of which indicate that the responsibility for the biodiversity crisis might be apportioned differently—most visitors have walked away at this point, apprehending the problem as one of overpopulation rather than overconsumption. Thus, while the curators of the Hall of Biodiversity name the causes of this crisis as population growth, agriculture, and urbanization, the encounter, at least with this display, lays the blame incipiently on the first. All the other problems—overfishing and deforestation, the impacts on human health, and the anticipated mass extinction—seem to arise from this one root cause.

Of course, this is not a narrative without precedence. The neo-Malthusian traces in this exhibit have been enshrined in other sites of knowledge production about the environment; international NGOs, world summits on the environment, the UN, and scientists the world over have all named it as a key issue. But what lurks here is a story of a shared earth in peril where the impulse toward green imperialism becomes justifiable. If the earth and its resources are indeed finite, then its management becomes a cause for global intervention. For the most part, the "population problem" is not located in the North (where, as the exhibit rightly notes the majority of resource consumption takes place) but rather in the South. What might be seen and read are the tropes that define the Third World in the Western imagination: overpopulated, "backward," and too fertile. The Hall of Biodiversity's approach, while definitely more nuanced, can serve to reinforce these narratives.

Further, the museum seems to count "the native" not only as too fertile, but also as closer to nature, part of the same move to position them as before or outside of civilization. So, in "Transformation of the Biosphere," one of the impacts of ecological destruction named next to species extinction is cultural extinction. Here, as with the rainforest diorama, we find that Indigenous peoples are positioned as anterior to culture: their folkways are closer to nature than Western civilization. Thus, as one of the curators of the hall asserts: "People are not purely the villains of the biodiversity crisis: Many cultures—hunter–gatherers such as the BA-Aka (Pygmies) and the San (Bushmen) of Africa and the Yanomami of South America; as well as pastoralists, nomads, and the pre-high-technological cultures in all the habitable regions of the world—have witnessed the loss of their traditions. Even when the peoples themselves survive, their cultures are lost to extinction" (Eldredge 1998, xi). So, while simultaneously imagining the bad Third World body, we have a definition of what the good one looks like: connected to nature, pre-technological, and close to extinction. This, of course, effaces the complexity of the Global South, emphasizing peasants and subsistence lifeways while ignoring the huge urban base that is now a reality of much of what was once called the Third World. It also categorically ignores the actions of some of its main funders, who contribute to the kind of "cultural extinction" that this part of the Hall of Biodiversity describes. Rather, the "native" becomes yet another part of nature that needs saving.

"Solutions"

The final element of the Hall of Biodiversity addresses possible remedies to the environmental crisis. The wall entitled "Solutions" is dominated by huge plaques devoted to the things that "we" can do to become better environmental citizens. Laid out in easy to read charts, this part of the exhibit explores five areas where citizens can get involved in the saving of biodiversity: protection and restoration, management for biodiversity, laws and regulations, reducing resource demand, and research and outreach. Each of these areas is then subdivided into strategies for learning about, getting involved in, and leading the way for biodiversity management. Further, different strategies are provided for different kinds of subject–positions: citizens, consumers, teachers, students, and parents. So, for example, museum visitors are encouraged to travel and consume responsibly, get involved locally and organize for enforcement of environmental regulation, "adopt a simpler lifestyle," plan their families, and "find your place in nature." In each section, the importance of science is underscored, and all subject–positions are encouraged to support the furthering of scientific knowledge so the impact of humans on the earth can be measured, assessed, and changed.

One of the more complex ways that life is managed through the museum is by its emphasis on technologies of the self: the conscious management of one's own conduct. In the work on governmentality, one key aspect of the shift in power was to emphasize the subject, or more properly the adoption of certain kinds of subjectivities over others to achieve the best end of the state (Scott 1999). Foucault articulated how this functions more broadly, indicating that power works "through progressively finer channels, gaining access to individuals themselves, to their bodies, their gestures and all their daily actions" (1980, 151–52). In the Hall of Biodiversity, there is an imagined objectivity in the way information is presented to the visitor—the knowledge is laid bare for the taking. Through this display, the museum visitor is presented with many options, from which they freely choose what to do with this newfound insight: "Basically what this museum does and even this exhibit is doing is to get information out there, to state some science, and make it so that people can understand it. And then they need to make their decisions based on that. I think that's the stance this institution takes, that they want to provide scientific information, and we are pretty careful about not overstepping the line with advocacy" (anonymous interview 1, 2006). But what is actually occurring is a highly prescriptive account of how humans should interact with and understand the environment, which produces certain kinds of actions/bodies/practices, rather than others. Again, the notion of environmentalism as religion comes to the fore, where adherents are asked to perform as ascetics, denying urges to consume in favor of the earth. Rather than providing a sustained critique of industrial capitalism (though these elements are in the exhibit to be sure, particularly in its discussion of deforestation and overfishing) the "Solutions" section of the exhibit mandates that the museum's visitors must look to themselves—their bodies, their practices, and the practices of the "other"—to find the remedies to the "Sixth Extinction."

Now, it is not to say that I think any of these strategies are inherently bad; indeed, I have incorporated many of them into my own life as I seek to mediate my own role in environmental destruction. However, through these technologies of the self, museum visitors are asked to imagine themselves in particular ways that are imbued with the pleasure of rightness (or righteousness). This is a very personal way of asserting truth. It functions for the most part by the very consumption choices that we make and implies that these choices are unconstrained—freely accessible for all to adopt. And we can feel good about ourselves when we choose to perform these remedies. However, not all people can inhabit and perform these subjectivities with ease. Indeed, these are particularly Northern (and within that categorization also presumes a bourgeois, autonomous subject) technologies of the self, that imagine a citizenry equally equipped to pay more for consumer products, engage in citizen action without repercussion, or have time to "appreciate biodiversity." Further, this making of innocence in the Hall of Biodiversity is compounded by the ease with which we can become these more virtuous kinds of people—by simply organizing locally, traveling responsibly, or consuming less energy. Like Disney's Environmentality Program discussed in chapter 2, the ecotourism espoused in chapter 3, or the solutions provided by *An Inconvenient Truth* in chapter 4, we are not called upon to change global structures of inequality that in large part produce ecological destruction. Rather, the answer lies in making simple changes in our day-to-day lives.

Practices of self-improvement are also urged by other aspects of the exhibit and its related materials. For example, the teacher's guide associated with the Hall of Biodiversity works to generate "good" scientific and ecological citizens through the promotion of natural observation for students. With activities like finding biodiversity in the backyard, keeping an observation journal, and making cladograms similar to that portrayed in the "Spectrum of Life," students in elementary and high school are encouraged to become budding naturalists: recognizing the world around us as full of nature in need of preservation. Further, there is a handout on resource use with an activity entitled "You've got the whole world in your shopping bag!" which seeks to train students to become better consumers. This, combined with the "Living with Nature: Consumer Choices for Children" program initiated by the Center for Biodiversity and Conservation at the museum, strikes at children early to teach them what are the better and worse ways of consuming the world.

Consuming Governed Nature

While the establishment of a particular form of governmentality is what this exhibit attempts, governing does not always operate as it is intended. Despite the nuanced operation of power, slippages, transgressions, instances of misrule, and instabilities always exist; or, put more succinctly, governing is never a completed project (O'Malley, Weir, and Shearing 1997). This is precisely because the museum operates as a space of encounter rather than transmission. In museum spaces, there are visitors who refuse to conform to the ordered nature of display, ambiguities in exhibits, and curators with a different vision. In this final section, I would like to address how these interstitial slippages occur in the consumption of the Hall of Biodiversity's presentation of nature.

The curators and others involved in mounting this exhibit intended to produce certain kinds of wants in museumgoers. They aimed to provoke understanding about the diversity of nature and hoped to inspire visitors to involve themselves in its preservation: "We want people to come to the Hall of Biodiversity and, without reading a word, say to themselves: How beautiful, rich and diverse life is! Do you mean to say this is all under threat? What can we do about this? Beauty is the key" (anonymous interview 2, 2006). Employing technology to make science understandable and grabbing viewers' attention with life-sized dioramas of tropical nature (rather than the boring old nature of New York City), the exhibit's designers hoped to encourage people to value nature in all its variety (interview 2). Moreover given the visitor feedback on a previous temporary exhibition entitled "Endangered," which mapped the plight of species in decline across the world, the staff of the American Museum realized that a positive spin, with some elements of hope (and a little bit of hype), was necessary to inspire change (anonymous interview 1, 2006). All of these insights went into the framing of the exhibit as a means to explore biodiversity, highlight its crisis, and provide concrete ways that people could make changes in their daily lives to forestall the "Sixth Extinction."

But, despite the efforts of museum staff, people rarely consume knowledge exactly as they are meant to. Indeed, as Bennett notes, "visitors practice their own forms of often quite unpredictable agency within the museum space" (2005, 528). In this vein, my tour through the exhibit revealed a myriad of ways of understanding the information presented. Some school groups quickly dashed through the Hall of Biodiversity—constructed as a hallway for maximum exposure—on the way to see the giant whale in the Milstein Hall of Ocean Life. Some lingered, reading all the supplied material and copiously taking notes. Others read the site as a spectacle, picking and choosing the more compelling aspects (such as the stuffed tiger in the "Crisis Zone") to examine and comment on. Parents directed their children's attention to elements of the exhibit they thought were the most important. Some children were so excited by the Dzanga-Sangha rainforest that they kept setting off the alarm that sounded when someone got too close to part of the exhibit. Some adult visitors were entranced by the same diorama, marveling in its intricacy and depiction of life in a place far away. Still others spent more time looking at the "Spectrum of Life" and less on the more text-based "Solutions" wall. All seemed to pay more attention when they noticed that I was taking notes on the exhibit, as if there was something more important to see since I was writing it down. However, quite a number of people expressed dismay at the information they were presented. One couple examined the Dzanga-Sangha diorama and pronounced that the shift from pristine to degraded nature was "depressing." Another couple made a similar comment when exploring the section "Transformation of the Biosphere": "This is sad." Then adding, "Do you want to go for lunch?" If we understand the museum as a space of encounter, then the visitors clearly shape and change the meanings offered by the curatorial staff. By focusing on the spectacle generated by the "Spectrum of Life," for example, one might leave the exhibit not with a sense of panic about the disappearance of such beautiful biodiversity but rather a feeling that such diversity overwhelms our capacity to change it. Some might just read the hall as a cabinet of curiosities instead of an unveiling of the cosmic order of nature. Questions about race, gender, class, and sexuality figure large here as well. How does a microphysics of power target people in distinct ways based on difference? How do bodies marked by difference encounter this space of governmentality differently? How do they read these spaces in complex ways? In the governmentality that the exhibit provides, there seems to be a particular blindness to difference that so often presumes whiteness as the baseline category through which information is displayed. For example, a person afforded all the privileges of whiteness and American citizenship might accept the storyline of integrated conservation and development projects in a way that a person of color or someone from the Global South might contest. The narrative of the intrepid scientist might be more easily read as masculinist by some rather than others. Indeed, the entire focus on reproduction and extinction of animal species speaks to a numbers game that can reduce animal life to a quest for, or blockage from, heteronormative coupling. And the consistent focus on consumption patterns as the tool for environmental change—through ecotourism or buying green products—clearly speaks to a particular class subjectivity that different people might consume quite differently. And so, the governing of the subject at the Hall of Biodiversity seems to have a very particular subject in mind; or at the very least, the person most likely to fully engage with the message is a white, bourgeois American.

And what remains unclear is whether the intention to provoke people to act—to incorporate the technologies of the self so elaborately placed throughout the exhibit—actually worked, an uncertainty that runs throughout this book. Despite these important breaches, however, museums are sites of power, and indeed the museum is always already a problematic space of (often forcible) collection, ordering, and display. What natural history museums do is present a compelling narrative of how nature should be understood. They operate as a discursive agent in producing the reality and "truth" of nature. It is because they act as knowledge/power formations that they are key to the production of green governmentality.

Conclusion

I think we can say that the Hall of Biodiversity at the American Museum is many things simultaneously. At the most surface level, it is a site of education (or perhaps now edutainment), a space where people come to learn about the past, present, and future of the natural world. Alternatively, one could agree with Luke (2002b) that the museum is a necropolis—a repository of millions of dead things, which are then put on display for human consumption and amusement. Using the language of poststructuralist theory, one could comment on the "heterochronic" (Foucault 1998, 192) nature of the exhibit, which seeks to enframe 3.5 billion years of evolution in one wall of the Hall of Biodiversity. I would argue that the hall is all of these things. But I assert that despite these different readings, a very particular story is memorialized here. Relying on both the modern episteme to reveal the invisible characteristics of life, as well as elements of the classical to organize nature into discrete and knowable characteristics and forms, this exhibit acts as an agent in constructing a certain kind of nature: a scientized assessment of global nature under threat.

To echo Dean's (1999) explication of the analytics of government, the Hall of Biodiversity does four things. It provides a characteristic way of seeing, rendering the causes of the biodiversity crisis visible to power and generating techniques for their solution. It maps nature through the biopolitical mechanisms of counting, ordering, examination, and display, rendering some populations (human and nonhuman) endangered and in need of management. It reinforces a subtle but present impulse to re-manage the nature of the South and save it from its inhabitants. And it speaks to the construction of certain kinds of subjectivities: the normalized green citizen imbued with caring for nature on the one hand, and the overly fecund body of the South on the other.

Of course, this does not mean the Hall of Biodiversity is a site of social control or repressive power. Rather, it produces. It produces knowledges; mechanisms for seeing; scientific expertise; techniques for classification, measurement, ordering, and display; possibilities for mediating risk; and subjectivities that embrace biotic life. It produces myriad maps, graphs, cladograms, films, reenactments, representations, and collages of both destruction and possibility. Indeed, as Foucault notes, the ability to produce knowledge is power's greatest feat: "[P]ower would be a fragile thing if its only function were to repress, if it worked only through the mode of censorship, exclusion, blockage and repression, in the manner of a great Superego, exercising itself only in a negative way. If, on the contrary, power is strong this is because, as we are beginning to realise, it produces effects at the level of desire—and also at the level of knowledge. Far from preventing knowledge, power produces it" (1980, 59). What the American Museum produces is ordered knowledge—heterotopic in form—where one encounters "a different real space as perfect, as meticulous, as well-arranged as ours is disorganized, badly arranged, and muddled" (Foucault 1998, 184). The hall brings classification, visibility, and understanding to the diversity and chaos of biotic life. Perhaps what the Hall of Biodiversity does best is establish a grid of intelligibility through which we can understand the nonhuman world. It establishes the vocabulary through which we describe biodiversity, and a system of signs to apprehend it. Thus, while it might not provoke immediate action from its visitors, it does organize thought. And in doing so, it seeks to govern the ways we can understand and act upon nature.

What I would like to suggest is that the kind of government at the Hall of Biodiversity in the American Museum of Natural History is a version of green governmentality with science as its primary scaffolding. Scientific representations, calculations, assessments, and predictions are given extraordinary weight in our society, and the sciences are understood as one venue, or perhaps the venue, through which truth is always told. At the same time, however, some of the science on display at this exhibit is popularized for easy consumption. Instead of staying with complexity, it reduces it so that its narratives can be easily understood by a wide variety of people. This, of course, makes sense given the numbers of people who will encounter the exhibit. But it also makes sense in terms of green governmentality, where the complexities of, say, conservation projects or the drivers of environmental destruction lead to a less coherent narrative. Yet, this "science lite" is complicated by the instances of science writ large, like the "Spectrum of Life" with its myriad layers of interconnection that are rendered visible for the visitor. In both instances, however, the exhibit works through the already legitimate and authoritative character of science, combined with technologies of vision and display that the museum has perfected through its many years of producing exhibitions and dioramas. In doing so, the Hall of Biodiversity can be understood as the manifestation of a power/knowledge nexus, where visitors are encouraged to consume the truth of nature as orderly and arranged, knowable and measurable.

The trouble, though, is that scientific fact and practice doesn't spring up preformed. Science studies scholars like Haraway (1989, 1991) and Latour (1993) have shown that scientific principles are made as well as found. The sciences, like all else, are always implicated in power; rather than simply describing the natural world, the sciences are enmeshed in the social, cultural, historical, political, technological, and economic structures in which their research and "discoveries" take place. And so, through science, the environment is brought into being, becomes an object of analysis, and its management then is a key aspect of governance.

Moreover, in this bringing into being, nature appears to correspond to a kind of order, a rational and coherent system that can be accessed and understood. However, nature (or culture or nature–cultures, for that matter) is not orderly. Categorizing it as such neglects the material dimensions of nature in favor of a more narratable discursive representation. It also forecloses, as was signaled in the discussion on the funders of the exhibit, other ways of encountering nature.

By way of ending this chapter, I'd like to point to a different way of knowing that points to the erasures that are present in this scientized view of nature under threat. Julie Cruikshank's (2005) work on stories about, and interactions with, the glaciers of the Saint Elias Mountains offers an example of the entanglements of nature and culture, science and folklore. She charts the conflict and connection between First Nations and European understandings of glaciers during the Little Ice Age between 1550 and 1900. The Tlingit and Athapaskan peoples viewed glaciers as sentient, reading the receding and growing of these icy mountains as the result of human/nonhuman interactions. The stories told about these interactions, involving the consequences inverting cultural norms, killing animals that make the glaciers home, and cooking with grease, or the capacity of glaciers to respond to acts of selflessness and reciprocity (see Cruikshank 2005, chap. 2), demonstrate that, for the Indigenous peoples who call this place home, "the categories of nature and culture were not the salient ones" (Haraway 1992, 236). In the stories that Cruikshank "listens for," glaciers have agency, they are "willful, sometimes capricious, easily excited by human intemperance but equally placated by quick-witted human responses. Glaciers engage all the senses" (2005, 8). Cruikshank places these stories in opposition to different ones: the local knowledge (though generally not understood as such) of the European scientists and American adventurers who traveled to the Saint Elias range. These alternate narratives foreclosed the possibility of the liveliness of nature, instead placing glaciers wholly on the side of an inanimate nature to be observed, its characteristics recorded for the scientific endeavor. Of course, the collision of these two ways of understanding nature was not separate from power; the complex webs of colonialism, science, nature, and culture described by Cruikshank have meant that the scientific understanding of glaciers has gained force. The local knowledge of European science has claimed universality.

The case of the glaciers of the Saint Elias Mountains provides examples where knowledge has been subjugated—been disqualified—by the weight of scientific discourse. Like the lessons provided by the Chinese encyclopedia, these knowledges have been buried, rendered irrelevant and parochial. But while they have been occluded, this does not suggest that these knowledges have ceased to exist. In an analysis of the continuing import of oral traditions, Cruikshank notes that Angela Sidney's narrative (one of the Tlingit women who recounted her life story to Cruikshank) works not only to retell stories passed down from her ancestors; her knowledge also has present-day relevance: "In her role as social commentator, she showed that while oral tradition can indeed illuminate the past, its meanings are not exhausted with reference to the past. Orally narrated stories also provide an alternate set of principles for thinking about current events" (2005, 57). If we take to heart the idea of a "land that listens" (47) provided by these stories, then there is a kind of reciprocity and relationship between the nonhuman and human that offers a potentially fertile space for difference. The stories that these subjugated knowledges provide exceed the binaries between nature and culture, calling forth a recognition of the co-constitution of the human and the nonhuman, staying with the complexities and the fleshy lived realities that a scientized green governmentality, the kind on view at the Hall of Biodiversity, elides. And it is here that we must remain if we are to think of a space outside of green governmentality.

CHAPTER 2

Disney's Animal Kingdom:
The Wild That Never Was

I N 1992 MICHAEL SORKIN published a picture of the sky over Disney World in his edited volume entitled *Variations on a Theme Park: The The New American City and the End of Public Space*. He did so to illustrate that the sky was the only thing that could be pictured without violating the copyright that the Disney Corporation has placed on the park. To Sorkin, Disney World serves as an example of the "contraction of the space of freedom" (1992, 207) that has become commonplace in American cities like New York and Los Angeles. For my purposes, it has meant that any picture I took in my research at Disney's Animal Kingdom in Orlando, Florida, can never illustrate my assertions in this chapter. Indeed, those who have used Disney images without permission have often suffered the wrath of a corporation with almost limitless means (for examples, see Budd 2005).

It is not only the images that Disney protects, but also what its employees say about the corporation. When I attempted to secure an interview with the director of the Disney Wildlife Conservation Fund and the Director of Animal Programs at Walt Disney World Resort (a process that initially proceeded quite smoothly), I was sent a release to sign from Disney's legal department. If signed, I would agree to: (1) have Disney review and approve how the interview material was incorporated into my dissertation; (2) portray the theme park and Disney's Wildlife Conservation Fund in a "positive light"; (3) not disclose any confidential information that came to me as a result of the interviews without seeking permission; and (4) not use the interviews for anything other than my study unless I received written approval from Disney first. Of course, I could not agree to these stipulations and had to cancel the scheduled interview.

These examples of Disney's vigorous attempts to defend its proprietary interests, often via threats of litigation, probably come as no surprise. Disney's control over its brand—its movies, television channels, animated characters, theme park environs, merchandising, and so on—is fundamental to Disney's business strategy. By strictly managing how their products can be used, in what contexts, and to what effects, Disney attempts to govern the way people can come to know and consume all the goods, discursive and material, that Disney sells.

This chapter delves into one facet of this governing by considering how Disney packages and sells nature as part of its repertoire of goods at the Disney Animal Kingdom Theme Park. Always a leader in reimagining (or "Imagineering," to use Disney's lexicon) and commodifying the structuring narratives of Western culture, Disney entered a new terrain with the opening of the Animal Kingdom Theme Park in 1998. With the park's catch phrase "The wild was never this wild," Disney seeks to reimagine and (re)produce nature as a site of sanitized, controllable, "family-friendly," adventure-filled fun. And in doing so, the Animal Kingdom works to remake nature as commodity, shaping popular imaginations of how it can be understood. Drawing on the work of those who have considered the cultural impacts of theme parks, like Susan Davis (1997), Susan Willis (2005), and Scott Hermanson (2005), I argue that places like Disney's Animal Kingdom are powerful sites for disciplining how nature can be conceived. However, what differs in my approach is the assertion that this power is also productive. Disney not only attempts to harness the affective registers and playful dimensions generated by a connection with nature in the quest for ever-increasing profits, which of course it does quite well. Its efforts also produce: they authorize Disney as a legitimate scientific actor and corporate citizen, they generate particular knowledges of and about nature, they empower particular subjectivities that rely on neoliberal economics, and they have "truth" effects that range far beyond the boundaries of the Disney Animal Kingdom Theme Park. In the end, Disney operates as a biopolitical institution, an important example of a corporation that successfully governs how nature can be experienced as fantasy, play, and spectacle.

In order to flesh out this argument, this chapter proceeds as follows. First, to provide context, I address Disney's influence as a cultural producer. Next, I turn to Disney's longstanding project of reimagining nature, which has culminated most recently in the construction of the Animal Kingdom. I then embark on an extended examination of the park itself, which draws upon the discourse analysis I conducted at this site. Next, I explore Disney's efforts to remake itself as an agent of conservation through its aptly titled "Environmentality" program and projects like the Disney Wildlife Conservation Fund. I conclude this chapter with a discussion about Disney's role as a biopolitical institution that enframes nature, as well as drawing out some larger lessons about how one of the most influential corporations in the world acts as an agent in the construction of a particular brand of green governmentality.

Disney's Empire

The Walt Disney Corporation is, without doubt, a huge media and entertainment conglomerate, with revenues topping U.S. $36 billion in 2009 (Walt Disney Company 2009). In terms of film production, Disney owns Pixar Animation Studios, Touchstone, Hollywood Pictures, Miramax, Buena Vista International, and Buena Vista Home Entertainment, and has produced some of the top grossing films, like *The Lion King,* which earned almost $800 million worldwide. In the field of television, Disney owns ABC, which has aired some of the highest rated television shows, including *Lost, Grey's Anatomy,* and *Desperate Housewives.* In addition, the corporation operates Walt Disney Television International, the Disney Channel, and ESPN, with interests in other cable channels, like A&E and Lifetime. Disney also owns a number of music labels, including Hollywood Records, Hollywood Records Latin, Lyric, and Mammoth. They have transformed some of their classic movies, like the *Lion King, Beauty and the Beast,* and *The Hunchback of Notre Dame,* into theater productions. And, Disney has a huge consumer products division, which includes the very successful Disney Store as well as Jim Henson's Muppets. Further, Disney also maintains two cruise ships, with plans to add two more, and has initiated a new vacation program called "Adventures by Disney," which offers guided tours to the Europe, North America, and Latin America.[1]

But of course, while the entire entertainment agglomeration is important to the iteration and reiteration of Disney's cultural dominion, the theme parks have become key artifacts of Disney's definition of childhood, innocence, family, and fantasy. Disney currently has theme parks in Anaheim, Orlando, Tokyo, Paris, and Hong Kong. The Walt Disney World complex in Florida has four main parks: the Magic Kingdom, Epcot, Disney–MGM Studios, and the Animal Kingdom. In addition, it encompasses two water parks, and Downtown Disney and Disney's Boardwalk for shopping, nightlife, and dining, as well as Disney's Wide World of Sports. Walt Disney World in Orlando employs 58,000 "cast members" alone to render the kind of experience that Disney seeks to provide its guests. Indeed, "immersion" or total body and mind entertainment is what Disney argues improves the consumption of their brand: "Our research suggests that the 'immersion' in the world of Disney afforded by the parks is a vital means of engaging with our audiences around the world and a key driver of affinity for the Disney brand, stories, and characters" (Walt Disney Company 2006, 17). In marketing the brand, Disney has situated itself as a key actor in defining what North American (and increasingly worldwide) childhood fantasies are, or at least ought to be. In doing so, a visit to a Disney theme park has become a pilgrimage that many middle-class kids seem hard-pressed to live without; as Bryman notes, "It is almost as though a visit to a Disney theme park is a benchmark by which parents judge themselves (and perhaps are judged by their children) as parents" (1995, 95). Histories and fairy tales, sanitized of their more objectionable or uncomfortable elements, are retold at Disney theme parks and serve to shape how children and adults understand a variety of abstract cultural categories—nation, globe, childhood, the "normal" family, nature—while commodifying all the way. Thus Giroux comments:

> The "Wonderful World of Disney" is more than a logo, it signi-
> fies how the terrain of popular culture has become central to
> commodifying memory and rewriting narratives of national
> identity and global expansion. Disney's power and reach into
> popular culture combines an insouciant playfulness and the
> fantastic possibility of making childhood dreams come true, yet
> only through strict gender roles, an unexamined nationalism,
> and a notion of choice that is attached to the proliferation of
> commodities. (1993, 87)

And Disney realizes this mission through strict control of how their narratives can be consumed, understood, memorialized, and retold.

This control stems in part from the fact that Walt Disney World in Orlando serves as its own local government. Using its leverage as a huge corporation that would provide thousands of jobs, Disney inked a deal in 1967 to create the Reedy Creek Improvement District that gave the company more governmental power than local municipalities to determine its own fate. The driving rationale behind this deal was the construction of the Experimental Prototype Community of Tomorrow (EPCOT), which was envisioned as a high-tech community in need of "flexibility" to keep it in a continuous "state of becoming" (Foglesong 2001, 69). Putting aside the fact that Walt Disney never intended the site to have permanent residents (69), the proposal offered the Disney Corporation the opportunity for complete control over the production, management, and consumption of its theme parks. As Foglesong reflects, the Reedy Creek Improvement District deal gave Disney the kind of power that other developers dream of:

> For the Disney Co. got something special in coming to Florida: their own private government, a sort of Vatican with Mouse ears, with powers and immunities that exceed nearby Orlando's. The entertainment titan was authorized, among other things, to regulate land use, provide police and fire services, build roads, lay sewer lines, license the manufacture and sale of alcoholic beverages, even to build an airport and a nuclear power plant. (In fact, Disney never established the public police force, relying instead on over 800 private security guards; nor did they build an airport or nuclear power plant, though they retain authority in state law to do so.) To the envy of other developers, Disney also won immunity from building, zoning, and land-use regulations. Orange County officials cannot even send a building inspector to Disney property, and sheriff deputies are obliged to check in when they come on property and to avoid conspicuous display of their marked cars on such occasions. (5)

In light of these developments, it is no wonder that Disney is litigious when its brand is used without permission. This agreement was signed in perpetuity, so that there is no end in sight to Disney's ability to control and maintain its property, image, and visitors. Under the aegis of the Reedy Creek Improvement District, Disney has its own brand of private government.

Disney's Animal Kingdom Theme Park

However, Disney's ability to govern doesn't end with the building of roads or establishment of private security regimes. I argue that this government has extended in less legal and literal terms to the ways Disney reconfigures and reimagines nature at its newest U.S. theme park, the Disney Animal Kingdom. Of course, this is not Disney's first foray into representing the biophysical world. Beginning with *Bambi* in 1942 and continuing with the

True-Life Adventures series exemplified by films like *Seal Island* (1948) and *White Wilderness* (1953), Disney was a pioneer in the development of nature drama for mass visual consumption.[2] As Margaret King notes, Disney reformulated an old narrative with regard to animals, where instead of being threatening or inscrutable, their stories became akin to human dramas, played out via notions of "anthropomorphism, selective perception, mixed motifs of pet/wild animal, the child/dog team, the cuteness/violence dualism, and a heavily edited version of natural events and processes" (1996, 62). Through editing, Disney was able to make nature "come alive" in ways that it could not possibly otherwise. Time compression works to enframe nature, not as a set of slow biophysical processes but rather a dynamic, rapidly changing, and always exciting series of events, each in close succession to the next (62). Presented was not the story of the daily lives of animals; rather, it is the predator seeking prey, prey avoiding capture, or the behaviors of animals that can easily be read through the lens of human emotion. Indeed, these early nature dramas served as morality plays, and narrating nature served to govern the interrelation between animals and people. This quest to define the proper (read: bounded) relationship between the human and nonhuman remains one of Disney's aims, and is artfully deployed and circulated in the various attractions of the Animal Kingdom.

The Animal Kingdom sits comfortably within the thematic notions of anthropomorphism, drama, and the separation of nature and culture that were ever-present in Disney's nature films. But it also goes far beyond them. At the dedication of the site, then-CEO Michael Eisner stated: "Welcome to a kingdom of animals . . . real, ancient and imagined: a kingdom ruled by lions, dinosaurs and dragons; a kingdom of balance, harmony and survival; a kingdom we enter to share in the wonder, gaze at the beauty, thrill at the drama, and learn" (dedication plaque at entrance). Encompassing all animals—elephants, tigers, and macaws alongside dinosaurs and yetis—the Animal Kingdom attempts to enframe all possible faunal natures, Disneyfying them for consumption in one site. In essence, Disney claims to have established a grid of all animals: actual, extinct, and imagined. Reminiscent of Foucault's comments about natural history in *The Order of Things* (2004), as explored in chapter 1, Disney positions itself not only alongside the likes of Linnaeus and Buffon but also the American Museum, as a repository of all that is known and knowable about animals. Further, in this same dedication, guests to the site are invited to situate themselves within the drama of the animal world, in congruence with the True-Life Adventure series. But they are also meant to *learn* in particular, and often commodified, ways. I will address this in detail later on in this chapter. For now, let me turn to the site itself.

Opened on Earth Day 1998, Disney's Animal Kingdom represented the fruition of Walt Disney's dreams of live animals at Disneyland. Apparently, he had wanted to use real animals in the now famous ride Jungle Cruise, but they could not be trained to perform as he desired; much to Walt's dismay, animals had minds of their own. And so, the idea was scrapped and Disney relied on audio animatronics to tell the animals' stories.[3] But Michael Eisner recuperated the idea of live animals at Disney during his tenure as CEO and the park was born. Zoo meets theme park, safari meets environmental education center, Disney's Animal Kingdom is 500 acres in size, by far the largest park in the Walt Disney World complex. The Animal Kingdom employs 4,500 people, who work as performers, education officers, maintenance staff, animal keepers, landscapers, shop attendants, concession cashiers, and more. It houses and displays 1,700 animals of 250 different species, and it cares for approximately 4 million plants, some of which require daily attention to maintain their "African look" (Birnbaum 2005; WDW News 2005b). Disney's Animal Kingdom attracts approximately 9 million visitors per year, ranking ninth in terms of theme park attendance worldwide in 2009 (TEA and AECOM 2009). The vision for the park was fueled by the Walt Disney "Imagineers" who "crisscrossed the globe in search of the essential look of life in the wild, amassing more than 500,000 miles . . . a distance equal to circling the globe 20 times" (WDW News 2005c). The work of these Imagineers resulted in a theme park that is laid out in the traditional hub-and-spoke pattern of other Disney theme parks: you enter the theme park through the Oasis, move into the central area called Discovery Island, and then can branch off into any of the five separate worlds: DinoLand U.S.A., Camp Minnie-Mickey, Asia, Africa, and Rafiki's Planet Watch (see map below).[4]

Before I embark on my discussion of the different "lands" at this theme park in earnest, a few words must be said about who visits the Disney Animal Kingdom, or at the very least, who was there at the same time I was in late June 2006. This time of year marks the summer rush, and so all the parks at Walt Disney World were awash with visitors. I found that the guests were remarkably diverse, in some ways shattering my notions of Disney as a monolithic white space. Of course, many of the visitors appeared to have

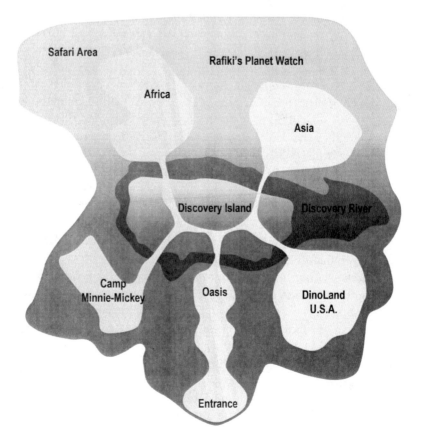

Map of Disney's Animal Kingdom theme park at Disney World in Orlando, Florida.

American accents, but British accents could also be heard, as well as other languages like French and Spanish. Many people seemed to be in large family groups that included grandparents, parents, brothers, sisters, and children.

There are two areas of common ground. First, the park assumes visitors have the means and subjectivity of the middle class. The first system of class exclusion that potential guests to Walt Disney World encounter is the cost of admission. The price of a one-day base ticket for Disney's Animal Kingdom is $87.33 for visitors age ten and up and $78.81 for children between three and nine. This, of course, does not take into account that many families that visit the park stay on the Disney property and buy weeklong passes rather than single tickets. In either scenario, however, the cost of fulfilling the desire for a Disney vacation can be quite expensive. Admission prices or hotel rates do not include, of course, the myriad gift items, souvenirs, refreshments, and food that are purchased as part of a visit to the theme park. Disney's Animal Kingdom, like other Disney parks, as many have remarked, is geared to the tastes of the bourgeois subject (see, for example, Bryman 1995; Fjellman 1992). Disney offers safe, clean, controlled family-centered fun, heavy on conservative moralism, and devoid of the stench of sex and violence. Disney's efforts at establishing a space that is ideal for the bourgeois subject has caused Bryman to remark: "The parks provided an image of Utopia that is not only congruent with middle-class values; the Utopia *is* middle-class America" (1995, 95).

The utopia espoused by Disney is not only middle class but is, perhaps unsurprisingly, also white. Many scholars have taken up how whiteness is made into an unmarked category, one that is normal, natural, and universal (Frankenberg 1997; Kobayshi and Peake 2000; Omi and Winant 1994). This conflation of white and normal in many times and places in the United States has meant that notions of what it means to be an American is often inextricable from whiteness. Of course, Frankenberg (1997) rightly asserts that this is not monolithic. She gives the examples of urban centers or border sites, for example, where the discourse of American as white is contested and, as a result, less coherent (5). However, Disney is not a cosmopolitan New York or a racially charged Atlanta, but rather a controlled space meant to replicate small town USA, a fantasy of whiteness. For example, people of color in the Animal Kingdom are made into hyperreal others to be consumed alongside the animals as part of the entertainment, a notion I explore in more detail below. This hyperreality is juxtaposed to the guest who, interestingly, regardless of race, is positioned as a person who can experience everything from a privileged position of invisibility. More succinctly, the guest is encouraged to perform the white gaze.

Linked to this notion of Disney's Animal Kingdom as a white bourgeois space is the fact that it is also a heteronormative one. And yet, this is complicated by the rise of "Gay Days" at Disney parks. Initially conceived as a radical act of gay activism in 1978, in the intervening years Gay Days have morphed into a fundraiser for AIDS (1980s) and later, an "apolitical vacation" (Griffin 2005). The notion of Disney World as straight space is further complicated by Eisner's estimates that in 1991 almost 40 percent of Disney "cast members" were gay or lesbian (131). However, neither Gay

Days nor the makeup of the Disney staff change the fact that Disney World is, in most ways, a site of compulsory heterosexuality. As "The Project on Disney" (a collective of authors who write under this name) has indicated, Disney theme parks are about "confirming family" of a particular kind (1995, 49). And indeed, nary a person was seen without her or his requisite child or grandchild during my visit to Disney's Animal Kingdom. In fact, this rendered my position as single woman a bit suspect, making my fellow guests wary of me on backstage tours. For Disney, family is understood as a heterosexual couple, usually with two children (Griffin 2005). The vast majority of people in the park meet this requirement.

The Oasis

The Oasis is the first area you encounter after riding the golf-cart trolley that takes you from the parking lot to the entrance. As the official Disney guide, *Birnbaum's Walt Disney World,* informs us, we are to imagine this crossing from the car to the gates not as walking across an asphalt vista so familiar to suburban America, but rather as the passage between northern and sub-Saharan Africa:

> Traditionally, one has to travel across a long, sunbaked stretch of desert in order to experience the soothing atmosphere of a tropical oasis. With that in mind, think of the Animal Kingdom parking lot as a concrete version of the Sahara. Once you've trekked across it, your journey takes you through the park's front gate and entrance plaza. . . . The transition is by no means a subtle one. Guests are immediately enveloped in a world of nature. The peaceful setting is most idyllic. (2005, 179)

Baudrillard has rather a different view of things—he names the parking lot outside of Disneyland, which might be extended to include the parking lots of Disney World, "a veritable concentration camp" (1994, 12). My view is somewhat more circumspect. I read it as a sort of liminal space, where one is offered the opportunity to escape the anxieties of urbanity, crime, difference, and complication in favor of a natural utopia. No matter how one imagines the parking lot, the visitor to the Animal Kingdom is asked to make a mental and affective journey, to suspend disbelief, something they will be asked to do again and again, and imagine themselves as part of a tropical paradise far removed from their ordinary lives: a site of true nature.

Once in the park, the Oasis is meant to swathe the visitor in nature at its most pristine—Nature with a capital N. Indeed, on Disney's backstage tour, named Wild by Design, I was informed that the Oasis was meant to represent "Nature before Man." The area was originally designed as an untouched paradise with no signs indicating which way to go. However, this idea was scrapped because guests felt unsure about how to proceed— a testament to the degree to which inside the gates of these theme parks, visitors surrender the decisions to Disney. In any case, the Oasis serves as a place to welcome people into the nature of the Animal Kingdom. To fulfill this vision of a pristine tropical environment, the Oasis is replete with caves, waterfalls, and plants that seem to overtake the paths, allowing you to shed the trappings of civilization and enjoy nature at its imaginary zenith (Birnbaum 2005). It is the absent presence of the genuine rainforest, in this case, which makes the illusion of the Oasis work. As Michael Sorkin has remarked of Disneyland: "[T]he whole system is validated, though, by the fact that one has literally traveled, that one has, after all, chosen to go to Disneyland in lieu of any of the actual geographies represented. . . . One has preferred the simulation to the reality. For millions of visitors, Disneyland is just like the world, only better" (1992, 216). The same is true in the Oasis and the other "worlds" of the Animal Kingdom, where enjoying a simulation of each place is cleaner, safer, and more appropriately commodified than the original.

The focus on safety, so key to the establishment of a bourgeois space, is evidenced through the Animal Kingdom's reliance on different kinds of surveillance. As you enter any Walt Disney World Theme Park, after your bag has been searched, you encounter a machine into which you are asked to stick your fingers. This machine measures the size of your fingers and attaches this measurement to your entrance card, giving you a biometric signature, which Disney argues, foils the would-be thieves of these often very expensive passes. What Disney produces here is what Haggerty and Ericson (2000) have called "data doubles": they have rendered their guests into information flows that, if linked up to other digital information on the guest, can be combined and examined at a later date. There has been some concern about what Disney does with the information it collects in this way, to which Disney has replied that it deletes the information thirty days after the card expires. However, it is interesting to note that Disney is the largest user of this kind of technology in the United States and indeed, after September 11, 2001, the U.S. government approached Disney for "advice in intelligence, security and biometrics" (Harmel 2006). The connections between Disney and U.S. intelligence do not end there, as former Disney executives have joined the National Security Agency (ibid.). I mention this to underline the governmental and governing power of Disney, and its multiple linkages to the state. Indeed, some have suggested, tongue in cheek, that because Disney is so good at biometrics, it should be made responsible for U.S. immigration. In any case, it is no surprise that Disney is a disciplinary space, where employees, guests, and animals are directed in particular ways. I will pick up this thread later in the chapter.

Discovery Island

From the Oasis, all paths lead to an area called Discovery Island, a pastiche of African and South Pacific architecture that renders it a feeling rather than a recognizable place. This site hosts the majority of shopping and dining establishments and at the height of the day is chock-full of visitors eating, drinking (nonalcoholic beverages), and buying in a sort of family-centered, heteronormative spree. Given that it serves as a point of convergence (all of the spokes emanate from it), it makes sense that Discovery Island hosts the focal point of the park, the Tree of Life. The tree is set amid lagoons with islands of wildlife—like flamingos, lemurs, and unlikely looking capybaras (a large rodent)—each with their own catchy poem and selected for their "beauty and playful behavior" (Malmberg 1998, 62). Rising out of this idyllic surround, the Tree of Life is a 145-foot sculpture of a tree that is meant to symbolize the interconnectedness of all life, one of the main motifs of the park as espoused by Disney. On its trunk there are carvings of more than three hundred different animals: a tiger, lion, elephant, rhinoceros, and bison to be sure, but also less mediagenic animals like a seahorse, octopus, warthog, and snake. Visitors gravitate toward it, marveling at the intricacy of the carving and playing a game to see how many animals they can spot. Typical of Disney's efforts at cultural appropriation (see the next section on *Pocahontas* and *The Lion King* shows), many of the artists who carved the Tree of Life were Native American, "because of their feeling for animals" (Hormay, qtd. in Malmberg 1998, 120). The tree is constructed out of concrete poured and sculpted over metal cages and filled out by 102,583 leaves, each a foot long, and made of plastic (120). It weds technological mastery and romantic nature in every branch. As Joe Rohde, Imagineer and Executive Designer for Disney's Animal Kingdom remarked: "The Tree of Life is a technological marvel, but it's also a symbol of the beauty and diversity and the grandeur of our animal life on Earth. . . . It's a celebration of our emotions about animals and their habitats" (WDW News 2005d). In some ways comparable to the "Spectrum of Life" at the American Museum, the Tree of Life stands as a visual representation of nature by Disney, knowingly consumed as a simulacrum, but better than anything nature could produce. But unlike the museum, Disney employs fantasy to narrate natural history. Released from the confines of scientific representation, Disney can create something of a hybrid artifact: both meant to delight in its artificiality but also to encourage a sense of interconnection between the various animals of the biophysical world. In this way, the Tree of Life is simultaneously fact and fiction, work and play, knowledge and pleasure. Edutainment on display.

Along the same lines, housed within the trunk of the Tree of Life is the animated 3-D movie entitled *It's Tough to Be a Bug,* which is meant to explore the world from an insect's perspective. The ostensible purpose of this show is to teach the viewers that bugs are beneficial to us: honeybees pollinate, dung beetles collect waste, spiders kill pests, and so on. The moral of the story, in another interesting link to the work of the American Museum of Natural History, is that insects provide essential ecosystem services for humans. It seems, then, that here again we find the focus on animals being necessary for humans: in terms of either our survival, or so we can admire their beauty. Of course, animals do provide essential ecosystem services. But this can also elide the intrinsic value of insects and other animals: that they have a right to exist in and for themselves. In their valuation as necessary and important actors, they are rendered without agency. What is interesting about this attraction is that, at least for small children, the 3-D film does not achieve its intended aim—in short, it doesn't govern. Although I read the warning that small children might find this film and its effects frightening, I was not prepared for the actual reaction the movie received. The child next to me was absolutely terrified and howled throughout the entire film. This, no doubt, had something to do with 3-D spiders and other insects shooting water and smoke out at the crowd. The delivery here lacks the edutainment capacity of other Disney attractions, and most parents left trying to comfort their small children.

Camp Minnie-Mickey

I begin the discussion of the main areas in Disney's Animal Kingdom with the "land" that is not really a "land": Camp Minnie-Mickey. The backstory to this area is that it is the summer camp where the Disney characters go to relax (Birnbaum 2005). However, it seems only to serve the purpose of separating the Disney characters (Mickey, Minnie, Goofy, and so on) from the live animals on display. Scott Hermanson (2005) suggests that this is because Disney needs to avoid an encounter between the fantasy and the real thing. However, the Animal Kingdom is a place that relies on the melding of fantasy and reality to produce a distinct version of conservation discourse. So instead, Camp Minnie-Mickey seems more akin to the segmentation of space and the creation of specialized sites, which is a hallmark of a disciplinary institution. Camp Minnie-Mickey has two main attractions, both of which rely on the popularity of their respective films: *The Festival of the Lion King* and *Pocahontas and Her Forest Friends*.

On the advice of the woman sitting next to me on the plane, an Orlando resident for twenty-seven years, I headed straight to *The Festival of the Lion King* show, which she told me was the most popular and "best" part of the Animal Kingdom. I was easily able to secure a seat and was then powerfully reminded of the kind of compulsory heterosexuality that is part-and-parcel of the Disney experience. Everyone—young and old—seemed to be in heterosexual pairings, usually with children. The show itself recapitulates many of the themes of the movie *The Lion King*, using singers and audio animatronics to entertain visitors and emphasize the importance of the "Circle of Life." The opening to what can only be called an extravaganza has four African American actor/singers onstage dressed in "traditional African garb": animal skins, brightly colored and patterned clothing, kente cloths, beadwork, and so on. The show is introduced as a "tribal celebration" where Kiube (Swahili for masculine and strong we are told) is the voice of authority, the leader of the group who wears a lion headdress. The other members are introduced as Nakawa (good looking), Kibibi (princess), and Zawadi (the gift). What follows is a "beastly pep rally," engaging the crowd in some hand clapping meant to boost their enthusiasm. Then the show starts in earnest with four audio animatronics emerging at each of the four corners of the theater: a warthog, a giraffe, an elephant, and of course Simba, the lion. They are flanked by acrobats also dressed in costumes that seem to be inspired both by "traditional" African dress and animals; skins, headdresses, monkey, and bird costumes pepper the performance. Then the show rehearses some of the famous songs from the movie: "I Just Can't Wait to Be King," "Be Prepared," "Can You Feel the Love Tonight," "The Circle of Life," and "Hakuna Matata." The performance ends with a rendition of "The Lion Sleeps Tonight," which was not part of the movie but nevertheless serves as a cultural marker to evoke Africa in the minds of the audience. What is interesting about *The Festival of the Lion King* is that, like the film, we are asked not to situate this place as an actual geographical location. Rather we know it is African because of the tropes, images, and music on which the show relies. Expertly timed and performed, *The Festival of the Lion King* brings together race, colonial imaginings, and the Western gaze in a short and happily consumable jamboree.

Similarly, the other attraction in Camp Minnie-Mickey is the show *Pocahontas and Her Forest Friends,* which draws upon the success of the Disney film *Pocahontas.* Leaving aside the film's reimagining of the colonial history of the United States, the sanitization of the story of Pocahontas, and the eroticization of empire that are depicted in the film,[5] the thirteen-minute performance does draw upon one element that suffuses the film: that Indigenous peoples have some kind of innate connection to nature. However, it expands this notion so that Pocahontas becomes the vehicle through which the "truth" about the need to conserve nature can be understood. In this performance, Pocahontas, amid the soundtrack of a chainsaw, struggles to find out why the forest is being destroyed. She notes, "it's [the forest] being cut down limb by limb, tree by tree.... They aren't even leaving a twig in their paths." She consults "Grandmother Willow," the talking tree from the film who acts as her advisor. Grandmother Willow reminds Pocahontas of the prophecy that says one of the forest's creatures has a special gift, a gift of protection and conservation. Pocahontas then talks to the animals: possum, porcupine, skunk, ducks, snake, raccoon—real animals, all of which are trained to perform to the delight of the viewing audience. Each of these animals teaches the viewers something about its ability to survive in the forest; each has its own special gift. But in the end, Pocahontas comes to the realization that the animal spoken about in the prophecy is a human; the only creature who can protect the forest is "us."

On the surface, the moral of this story does not seem such a terrible thing. Indeed, it works in part to remind viewers that conservation is important and that humans have done much to destroy habitats like, for example, the site where Disney stands, which was once a vibrant part of the

Everglades ecosystem. However, as Hermanson's analysis of *Pocahontas and Her Forest Friends* shows, this performance shores up the binary between nature and culture that is fundamental to Disney's Animal Kingdom: "Implicit in Pocahontas's lesson is the idea that nature is an absolute—a pure, unadulterated, authentic force that stands in contrast to the artificial, compromised world of humans" (2005, 204). This lesson is problematic in three ways. First, it imagines that nature is where humans are not, so we must work to preserve the last of a fading wilderness while nature in the in-between spaces—urban natures being the most obvious example—are relegated to the unnatural (see Cronon 1996). Second, it draws lines around what is defined as nature and culture, respectively. These divisions have particular consequences in the world. In this case, for example, it positions Aboriginal people as closer to nature (and because of this more responsible to alert us to its problems), evidenced by Pocahontas's ability to speak both to animals and tree spirits in the form of Grandmother Willow, and "other" people—those who are destroying the forest—as outside of nature's thrall. Simultaneously, however, while defining the division between people, the parable afforded by *Pocahontas and Her Forest Friends* serves to recenter humans (monolithic, without difference or differential responsibility) as both destroyers and saviors of the biophysical world. In either case, nature exists only as we ruin or save it: it does not have agency or meaning outside of that which humans ascribe to it. In what could have been written specifically about this performance, Kay Anderson notes of bioparks that "[t]he animals enter the stage as monuments to their own disappearance in nature. This is the post-modern biopark of the 1990s that bears the stamp of an insistently human discourse—a globalizing narrative of nature's loss (at human hands) with the promise of its heroic, human-led, recovery" (1995, 290). It is this last point that becomes central to notions of nature in the next "land" I will discuss: DinoLand U.S.A.

DinoLand U.S.A.

Themed as U.S. culture at its gaudiest, DinoLand U.S.A. is a reproduction of "roadside America . . . the likes of which you might stumble upon during a cross-country road trip" (Birnbaum 2005, 186). Drawing on the mythology and actuality of America's obsession with the car, DinoLand U.S.A. invites park guests to experience a road trip without having traveled (though, no doubt, some visitors did engage in long family drives to get to the Animal Kingdom). So, you are treated to games, concession stands, and a ride that you might find at a state fair, done to Disney perfection. Partnered with McDonald's, DinoLand U.S.A. represents the union of the two corporations that are central to the production of childhood desire.

But DinoLand U.S.A. is not only about American kitsch. It is also meant to celebrate the supposed American fascination with dinosaurs that was born through the collecting expeditions of places like the American Museum of Natural History. Consequently, the visitor encounters large-scale reproductions of fossils, a spot to discover the linkages between dinosaurs and their descendents, and the Boneyard where kids can play at being paleontologists. The centerpiece, however, is a ride called DINOSAUR, which is housed within the Dino Institute, a fictionalized natural history museum crossed with an imaginary scientific research center. The storyline of DINOSAUR is that as part of the paleontological institute, the rider's mission is to go back in time to rescue the last Iguanodon and bring it back to the present to understand why dinosaur extinction occurred. As with many of the attractions at the Animal Kingdom, the wait in line is as heavily themed and entertaining as the ride itself. You enter a rotunda with a fossil named Dino Sue, a thirteen-foot replica of Sue, the most complete Tyrannosaurus Rex skeleton ever discovered. (Indeed, Disney, as well as McDonald's, supported the acquisition and exhibition of the real Sue at the Field Museum in Chicago [Kadanoff 2002].) In addition, you get a voice-over introduction by Bill Nye, the much-loved television "Science Guy." Following Dino Sue is an area where a video is played starring Phylicia Rashad (of *The Cosby Show* fame) as Dr. Helen Marsh and Wallace Langham as Dr. Grant Seeker, the coordinators of the mission, who stake out the parameters of your journey and impart riders with a sense of purpose. Once on the ride, the path is more predictable; the visitor encounters audio animatronic dinosaurs and the requisite thrills. And, of course, in the end the mission is completed successfully.

This ride had no tie to conservation that I could discern, but like the other, more conservation-themed attractions at Disney's Animal Kingdom, it does stress the importance of science. The rotunda replete with fossils (or more properly their replications), the notion of the Dino Institute with its requisite graduate students and professors, and the ultimate quest to find an Iguanadon before it can become extinct: each of these elements of the ride point to the way that Disney deploys an objective and legitimate science, and works on our assumptions about natural history museums, albeit in a playful way. Disney tells us "the future is truly in the past." But the past can only be accessed through science: technology can eliminate extinction—indeed, it can reach back and pluck species out of their space and time for examination, categorization, and conservation today. At DINO-SAUR, we are meant to understand that the science can change the course of natural history. If the quest is to bring back a dinosaur specimen to understand why and how extinction occurred, then certainly it is feasible to assume the end result would be to forestall that extinction. As Hermanson has rightly indicated: "One of the most potent arguments for environmental reform is the cry 'Extinction is forever.' Countdown to Extinction [now named DINOSAUR], while clearly a fantasy, plays on our belief that technology will eventually ride to the rescue" (2005, 217). And yet, one cannot go down this road too far. The ride is clearly understood as a fiction. While genetic harvesting is a reality and time travel a future possibility, none of the ride's participants actually imagines they are engaging in scientific discovery. Rather, it is the rehearsal of a potent narrative, fictionalized here but made fact in other sites, that makes the ride go, so to speak. In this move, one can locate its governing effects. It doesn't generate knowledge, but rather relies on well-known practices and performances of scientific pursuit to educate through both play and pleasure.

DinoLand U.S.A., along with Camp Minnie-Mickey, is meant to define Americanized nature. What is interesting, then, is that it pictures nature as dead, extinct, ghostly. Susan Willis argues that DinoLand U.S.A. serves as a narrative of eventuality, where the dinosaurs are exemplars for what is to come for all animals: "By devoting an entire subpark to the theme of extinction, Disney situates its wildlife and wild lands in a countdown to the inevitable . . . a parable of death" (2005, 67). While I find this argument intriguing, I am also not convinced that this is the meaning behind Dino-Land U.S.A. Instead, DinoLand U.S.A.'s focus on extinct animals, and how we can recapture them through archaeology or time travel, works to position America as having left the trappings of nature behind in favor of science. Here, visitors come to understand the salvational power of science to solve the environmental crisis. Not unlike the narrative of generating environmental services through harvesting biodiversity, DinoLand U.S.A. provides a possibility that the destructive impetus of modern (post-) industrial capitalism might lack permanence, given the right application of technological will. In a sense, guests are left with the notion that extinction is not forever—animals can be recuperated from the past with good science and good will. If this is the case, then one need not worry so much about conserving resources or reducing consumption: science will rescue us from our overindulgence. This erasure of nature in favor of science stands in stark opposition to the lands I will discuss next—Asia and Africa—which teem with the nature that the United States has eradicated.

Asia / Anandapur

In "Asia," or Anandapur, meaning "place of delight" in Sanskrit (Birnbaum 2005), Imagineers wanted to evoke the essence of Asia and distill it into signifiers that are easy to identify and digest. As such, this "land" was designed to be a composite of socio-natural markers from Nepal, India, Thailand, and Indonesia (WDW News 2006c). As I was informed on the Wild by Design backstage tour by one of our guides, although China or Japan might immediately come to mind when Asia is mentioned to the "average American," such places were far too industrialized to depict the sort of idyllic nature Disney wanted for its theme park. Instead, Disney embraced the trope of Asia as wild with nature, a place to be cherished as a site of pristine wilderness. The guide on the tour also informed me that the storyline of this part of the park is that locals have realized the importance of their nature and are promoting ecotourism as a sustainable way to preserve nature and "enter the world economy." This backstory is key to the one centerpiece of this land, Kali River Rapids (now perhaps eclipsed by the new thrill ride Expedition Everest), but is also evident in the concession booths located around "Asia," which are themed as ecotour operators that provide adventures for the willing Western tourist. Here in "Asia," capitalism meets conservation to good effect, at least according to Disney. Each one of the attractions in this section of the park—the Maharajah Jungle Trek, the bird show Flights of Wonder, the new roller coaster called Expedition Everest, and the Kali River Rapids water ride—take up notions of conservation, capitalism, science, and the Western gaze to tell the story of Asia, or at least how Asia should be. Corporate America imagines the Global South.

The Maharajah Jungle Trek is a self-guided walking tour, the purpose of which is to observe the animals of the Asian rainforest. This has more of a traditional zoo feel, where you wind your way through what is themed as an ancient ruin of a royal preserve that was opened up to the public to educate the locals about the importance of animals. Throughout the path, you encounter komodo dragons, tapirs, tigers, deer-like animals, and various birds. In the middle of the walk, you come upon a "research station" where you can observe vampire bats and look at reptiles encased in terrariums. The path then wanders through a sort of aviary and ends at a four-paneled bas-relief that depicts a parable of the human encounter with nature. The first panel shows nature before humans: paradisiacal, idyllic, lush. The second panel portrays a man destroying the forest by cutting down the trees. The third panel shows how this act unleashes havoc into the world, so much so that the man himself is threatened by his actions. The last panel indicates that the man has learned the error of his ways and has been redeemed, choosing to replant to renew the forest, though it is permanently affected and not all of the animals will return.

The Maharajah Jungle Trek is similar to the Cretaceous Trail in Dino-Land U.S.A. or the Pangani Forest Exploration Trail in "Africa" (discussed below) in that it reads more as the average zoo. It also shares with the Pangani Trail a theming of the importance of science, as evidenced by the "research stations" found in the centers of each walking tour. However, this four-paneled bas-relief makes the Maharajah Jungle Trek something different. Although not in a place where many of the guests might notice it (indeed, it appears at the end of the walking tour, where some teenagers were veritably running to escape what they felt was a boring attraction), in some ways this piece of sculpture sums up what Disney is attempting to portray in other parts of the park: "we" (and again this is an undifferentiated we) have abused nature but "we" also come equipped with the tools to remedy our mistakes. Almost all aspects of Disney's Animal Kingdom point to this discourse of sustainable development or the familiar narrative of "the world in the palm of our hands."

Moving on from the Maharajah Jungle Trek, the next attraction is "Flights of Wonder," a live show that incorporates trained birds into Disney's story of the necessity of conservation. In the show, there is a main character who acts as an educator, telling the audience about birds for a couple of minutes until a sort of bumbling tour guide/explorer named Joe Guano runs into the show and a variety of high jinks ensue. The main thrust is that these birds are both beautiful and useful and thus need to be preserved. For example, the audience is told that the African crane keeps down pest populations that destroy crops and carry diseases. However, despite the value of these birds, their habitat is threatened and "we" need to become involved to save them. How? The following dialogue is taken from the show:

JOE GUANO: The world is our greatest gift, and it's up to us to preserve it.

HOST: That's very well said folks. We as humans are the only species that can do that. It just takes simple things like conserving, recycling, supporting conservation efforts.

JOE GUANO: Oh you mean like Disney's Wildlife Conservation Fund at Disney's Animal Kingdom?

HOST: Absolutely, efforts like that can make a big difference for animals all over the world. . . . Guess that reminds me of something I once heard: "We haven't inherited this planet from our parents, we're borrowing it from our children."

JOE GUANO: Friends, the heart and spirit of all living creatures share a common connection.

Through this exchange we are reminded that only humans can save nature and that this can be done through sound environmental stewardship, made manifest in fairly simple techniques of recycling, giving money to conservation organizations, and conserving natural resources. This straightforward narrative seems to slip a bit at the end of the exchange where Joe Guano notes that, "all living creatures share a common connection," seeming to imply that humans and nonhumans are necessarily linked. However, even in common connection, it seems as though "we" are in charge. Moreover, this discussion positions Disney as a good corporate citizen with a heart and soul. Harder questions, like the training and performance of animals for human amusement are left, necessarily, untouched. Instead, our anxieties about species extinction are assuaged because, through simple means, often through curbing or changing our consumption patterns, conservation can be achieved.[6]

Moving to the next attraction in "Asia," Expedition Everest is the newest ride to be opened at the Animal Kingdom. The backstory of this ride draws upon the legend of the yeti in Nepal. The ride itself is themed as a quasiscientific journey on a mountain train to discover whether the yeti exists. However, midway into the ride the track has been destroyed by the yeti. The climax of this roller coaster occurs when you come face to face with the beast, the biggest audio-animatronic ever created by Disney, who serves as protector of the mountain. After the requisite roller coaster hills and thrills, the riders are returned to the base camp having survived their encounter with the mythic creature.

What is interesting about Expedition Everest is not so much the ride itself but the theming that surrounds it. First, it is set inside a reproduction of Mount Everest, which rises above Anandapur and serves as a focal point for the area. But, in addition to the mountain, the ride is also set within a fictional village titled Serka Zong, which, as a Disney press release states, is "like the mountain, a marvel of authentic detail. The village consists of several buildings, including a hotel, Internet café and trekking supply store, all reflective of today's Nepalese architecture. A canopy of prayer flags, an ornamental monastery, intricately carved totems, and a garden of stone carvings of the yeti clutching the mountain immerse guests in a far-off realm" (WDW Magic 2006). This search for "authenticity," for the correct detail in minutiae, works to evoke a sense of the place that is reductionist. Here, like in the *Festival of the Lion King,* we have the markers that might define Asia in the Western imagination—the external trappings that then stand in for a place itself.

But the theming does not end here. Importantly, the ride is also framed by the actual expedition that informed it: the scientific journey that Disney and Conservation International undertook to the Himalayas that was filmed by the Discovery Networks. The area where you wait in line is littered with memorabilia from the expedition, the quest for more knowledge about a far-flung ecosystem. This expedition, unlike others that Disney has engineered, was not wholly about mining for Imagineering fodder. Instead, it was framed as a kind of scientific trek. Before embarking on the expedition, a Disney press release noted:

> The team of internationally renowned biologists, botanists and technical experts will conduct a scientific inventory of plant and animal species in areas that are little-known but potentially important conservation sites. "Due to the fact that this region has gone largely unexplored, we believe that, in all likelihood, new species of plants and animals not yet known to science will be discovered," according to Dr. Russell Mittermeier, the world's foremost primatologist and president of Conservation International. (WDW News 2005a)

These are the kinds of assertions you expect to hear from the American Museum of Natural History rather than Disney. But in fact, the expedition lived up to this prediction. Walt Disney Imagineering and Conservation

International "documented a significant number of new, rare and endangered species" of beetles, amphibians, and insects hitherto unknown to science (ibid). And, of course, the entire journey was caught on tape by Discovery Networks and broadcast to the world via the Science Channel, the Travel Channel, the Discovery Channel, and Animal Planet (WDW News 2006c). As Mittermeier, already positioned as an expert in scientific inquiry, asserted: "We are thrilled with Disney's dedication to conservation through their scientific and financial contributions to the expedition" (ibid.). Here, a corporation takes on the role of museum and scientific institution, connecting its goal of building an attraction to the higher purpose of a nongovernmental organization.

The final attraction in "Asia" is the Kali River Rapids ride. This ride underlines the backstory of "Asia": ecotourism as a way to save nature while generating "good" profit. Kali River Rapids is supposed to be an ecotourist rafting adventure in this unnamed Asian place. As you wait in line for the ride, you are witness to various imaginings of Asian culture: an antique shop ("antiks made to order"), various stalls selling fruit and other wares, and beautiful murals depicting lessons from the animal world. Once on the ride, the linchpin is that as you float along through the rainforest on a rafting trip, you are confronted with an illegal logging operation where the forest is being burned. Unfortunately, the ecotour organizer is at the temple (like everyone else who is supposed to be operating the tour), and even though your guide radios the base camp, the boats cannot be stopped from coming. Perhaps this indicates that "resistance is futile"; that the onslaught of tourists, of development, of nature consumption cannot be stopped. In any case, the logging has been reported and we are to "make believe" that authorities will intervene. The ride proceeds down a waterfall, and all the "ecotourists" are returned to safety, allowing them to successfully forget about this issue.

The notion that ecotourism is the appropriate way to save nature, rescuing via consumption yet again, is paramount to this ride. Of course, ecotourism itself is contested and often read as an imperial endeavor. But for those who can afford it, tourists are offered an intimate experience with wilderness, a chance to encounter "authentic" cultures and possibly reinvent themselves in the process, all of which, of course, is the focus of chapter 3. However, in the context of the Animal Kingdom, ecotourism sets up a dynamic in which capitalism is the saving grace that nature needs. Indeed, if not for the presence of Western subjects, we are to imagine that this illegal logging could never be stopped. In turn, this permits guests to consume the other—imagined in this case as either well-intentioned but bumbling ecotour operators or agents of rapacious capitalism—while reinscribing their own identity: the enlightened tourist performing the "green man's burden" (Luke 1997, 35).

Hermanson (2005) has indicated that the story presented in the Kali River Rapids is a complicated one. At first glance, it seems to point to the avarice of capital and its role in destroying ecosystems. In this way, it might be read as an indictment of capitalism. However, Hermanson (2005) alerts us to a crucial part of this narrative that diffuses any possible critique: the logging company in question operates illegally. He argues that "by placing this particular company outside the law, Disney makes its corporate villains a bit more threatening to the rafters yet manages to avoid a more serious critique about normalcy in everyday corporate practice" (208–9). Instead, we are asked to imagine that only renegades, corporate or otherwise, engage in such activities. This notion of illegality is repeated in the next land examined: "Africa."

Africa/Harambe

Entering the "Africa" section of the park, one is once again struck by the lengths Disney went to replicate the markers that define Africa in the U.S. imagination. In Harambe, the fictional village brought to life by Disney, you encounter architecture, bazaars, and foliage that signal a change of continent from Asia to Africa; though again there is the sense that it is both anywhere and nowhere all at once. Built to resemble a coastal East African town, Birnbaum's guide informs guests that "[e]verything here is authentic. . . . After seven years of observing, filming, and photographing the real thing, they recreated it here in North America" (2005, 182). The Imagineers note, "[w]hile inspired by the town of Lamu in Kenya, Harambe designers chose not to copy a single street or marketplace but to capture the essence of the busy coastal city. They collected native artifacts, distinctive signs and designs right down to the cracks in the sidewalk" (WDW News n.d.). Thus, the buildings are suitably peeling, African music wafts from nowhere, trees are pruned to look as they do in East Africa, the cobblestone walkway is quaint in its unevenness, and safaris dominate the economy. Harambe is even replete with "cast members" in African-inspired costumes greeting visitors with "jambo," the Swahili word for hello, although one wonders what an African would make of an Africa so filled with the white faces of both staff and guests. In any case, Disney's Africa has two main attractions that I will explore in more detail: the Kilimanjaro Safari ride and the Pangani Forest Trail.

The Kilimanjaro Safari is one of the key attractions at Disney's Animal Kingdom, and is billed as Africa, only better. As the advertising for this ride on the Disney Channel in the hotel I was staying at indicated, it was Africa, just without the mosquitoes. In fact, the New York Bureau Chief of *The Economist* online noted that he "[found] himself preferring Disney's hyper-real version—especially as the appearances of the animals are so well choreographed that I start to suspect they may be robots, or next-generation animation" (Bishop 2007). In line, visitors are shown a video where a supposed African game park guide tells of the importance of conservation and the impact of illegal poaching on animals, as opposed, one imagines, to the controlled safari expedition. This primed us for the ride, which is, in part, about seeing the animals in a better-than-African experience of the Serengeti, but also about demonizing poachers and enrolling guests in the fight against them. Again, the visitor has no need to visit the actual place: Disney has delineated, replicated, and improved upon all its important parts and refashioned it for the consumption of American audiences, including potholes and falling bridges. Here nature has become gentrified. You travel along in a Land Rover–like vehicle and spot animals on either side. Giraffes, warthogs, watusi cattle, and cheetahs to the left; lions, kudu, alligators, and ostriches on the right. The landscape itself is made actor as well, with manufactured kopjies (a rocky outcrop), baobab trees, and acacias that give the ride its "oh so African" feel. However, the climax of this ride comes when guests hear over the radio that there are ivory poachers on the loose who have captured two elephants: audio animatronics Big Red and her daughter Little Red. The remainder of the ride then becomes a bumpy kind of Wild West movie set in Africa to catch the poachers before they can do damage. In the end, of course, the poachers are caught, the elephants are saved, and order is restored.

This ride is similar to the Kali River Rapids, both in terms of its reliance on illegality to construct the environmental villain and its reproduction of imperial forms of consuming the other. On Kilimanjaro Safari, riders are told that the issue that affects the African environment is poaching. The story is not made more complex; perhaps that poaching occurs due to a loss of land as conservation in some places equals enclosure. Nor is the guest witness to the myriad of other activities that threaten African wildlife, such as transnational mining operations, forestry, export-led monoculture, and the like. There is an easier alibi here, much like the illegal logging seen on the Kali River Rapids ride. By virtue of the unlawfulness of capital on both rides, other kinds of corporate endeavors are less implicated in the destruction of nature, if not made completely innocent.

Further, the kind of tourism portrayed on this ride, the African safari, is a complicated story. Susan Willis (2005, 54) has suggested that the Kilimanjaro Safari ride draws both from the late nineteenth- and early twentieth-century tradition of big game hunting and the adventure genre in movies, which conflicts with the environmentalist push for preservation. But do these ideas necessarily conflict? Is big game hunting or Indiana Jones, to use Willis's example, so far distanced from the dominant discourses of modern environmentalism? Indeed, it seems to me that the safari is the perfect representation of the way that different and often oppositional narratives can collide in global environmental politics. Consumer politics broadly and ecotourism specifically have become one space where the environmentally minded can express their values. Real-life ecotourism, much like this fictional safari ride, positions mostly upwardly mobile white tourists as the consumers of a disappearing nature, a nature that they can save through their very act of consumption. The safari is akin to the big game hunt, only with cameras rather than guns; the aim, if not the effect, remains the same: to capture the animal (Haraway 1989).

From the Kilimanjaro Safari ride, I moved onto the Pangani Forest Exploration Trail, which as mentioned above, is very similar to the Maharajah Jungle Trek, with the exception that it is themed as a conservation school (Birnbaum 2005, 183). The exploration trail is a winding self-guided path where guests can see a variety of different kinds of wildlife—meerkats, African birds, gorillas, hippos, and vultures. At each station, there is a docent who explains further about the animals and the need to preserve them. The exhibit is framed around a "research station" that lends further legitimacy to this site—actual research is being conducted on these real animals, with particular reference to Disney's work on animal hormones. Within this staged research station are exhibits of a kind: insects and rodents behind glass, ever available to the gaze of the viewer. The attraction reaches its climax at the observation window for the Gorilla Research Camp, a sort of greened lab. Here again, similar to Expedition Everest, science is presented as a key aspect of this attraction, and visitors can imagine themselves as research scientists.

The Disney versions of Africa and Asia share many similarities. The Imagineering witnessed in Expedition Everest, in the sense of producing the truth of a place, defines the Africa and Asia sections of the park. Of course, Disney has a long history of producing hyperreal spaces like Main Street USA, Tomorrowland, and Frontierland, which Baudrillard comments are "the perfect model of all the orders of simulacra" (1994, 12). Unlike Tomorrowland, however, Africa and Asia have a material reality. Drawing on the tropes of the Global South as humid, languid, lush, and perhaps just a little primitive, a very particular image of Africa and Asia is evoked. Relying on the fact that most of their visitors have never, or will never, go to these places they are eager to consume, Disney draws upon stereotypes to stand in for Asia and Africa. In this way, these sites operate as Baudrillard's simulacra, supplanting the original places and reinventing them as spaces of "neocolonial kitsch" (Hermanson 2005, 209). Indeed, the typical African safari after which the Kilimanjaro Safari ride is fashioned was judged to be less realistic than the copy; Africa has a quality of "theme park-ness" according to the Imagineers who visited it (Malmberg 1998, 15). What this means is that these places become the sum of the markers that are used to define them. They are notable in that they become spaces that can and should be represented for consumption, but simultaneously are expunged of the histories and agencies (human/non-human/technological) that made them. So, for example, the African savannah is understood as a site for struggle between humans and nature, defined by issues of population and poaching (15). Asia is imagined as an ecotourist haven, where the most pressing environmental issue is illegal logging. In each of these cases, what gets represented is a sanitized version of these places with easily consumable narratives. Much like the stories told in world fairs, these places do not talk about violence, epistemic or otherwise. (Neo)colonialism, slavery, resource extraction, poverty, accumulation by dispossession, and so on—actual histories in short—are silenced as they are not themes that are easily or happily consumable. Instead, these places are reimagined, embalmed, and made into artifacts. Disney works to fetishize particular parts of Africa and Asia, or to decontextualize and recontextualize to use Fjellman's (1992) words, and disregards those aspects that do not fit into the marketing package or challenge the values of the market.

Similarly, Disney works to compress not only space, but also time:

It [Disney] is not only interested in erasing the real by turning it into a three-dimensional virtual image with no depth, but it also seeks to erase time by synchronizing all the periods, all the cultures, in a single traveling motion, by juxtaposing them in a single scenario. Thus, it marks the beginning of real, punctual and unidimensional time, which is also without depth. No present, no past, no future, but an immediate synchronism of all the places and all the periods in a single atemporal virtuality. (Baudrillard 1996, 2)

This notion of the collapse of space and time, and their re-articulation in one site is yet another element of the world fair. Further evidenced here are elements of the commodity spectacle as discussed in chapter 1. Drawing on the example of the Great Exhibition of 1851, McClintock (1995) demonstrates an early example of time/space compression, where a panoptic imperial vision of the colonies was displayed. But, of course, this display was not neutral. The gathering of cultures seen first in the Great Exhibition of 1851, and replicated many times over in subsequent world fairs, served to make known national or territorial hierarchies. This is no different in the case of the Disney's Animal Kingdom. And so, we have DinoLand U.S.A., which stands in complete opposition to Africa and Asia. This "land" represents U.S. nature as dead, as the dinosaurs. In spite of, or perhaps because of, this fact, the United States is the site of all the other natures, and is the space where they can be assembled, judged, controlled, and consumed. Indeed, one might read the layout of the park as a map of the empire, with each "land" corresponding to different ideological frames of how the world is structured. While Africa and Asia teem with wildlife, DinoLand U.S.A. focuses on extinct animals, and how science can recapture them, or at least their DNA.

How can we reconcile these two views of nature, one dead and one living? My reading of these two representations is that they function to reinforce the nature/culture dualism so present in Western society. Thus, what we are meant to understand is that Africa and Asia, and their inhabitants, are closer to nature, while the United States is closer to civilization, having left the trappings of nature behind in favor of science. Progress is, in some sense, the organizing theme of the Animal Kingdom. Given that the United States is positioned as a leading nation of science and culture, this construction of U.S. nature as dead also serves to fix the other cultures represented at the Animal Kingdom in time. Thus, a visit to Africa or Asia is a trip back in time (or maybe an altruistic future), rooted in a prior era where nature was lush, fecund, and abundant. As time has moved on in the United States, however, so too has nature, leaving DinoLand U.S.A. rife with death. So, the imperative becomes to save that nature which has been fixed in time, which has not yet evolved on the linear path of modernity. Through ecotour adventures, safaris, and scientific research institutions, the natures of these anachronistic sites can be rescued from their eventual extinction through acts of consumption and capitalist investment. And, of course, this allows a whole host of other, often unconscious, distinctions to be drawn: primitive/civilized, natural/unnatural, us/them. In turn, this permits guests to consume the other—both human and nonhuman—while reinscribing their own identity. These sorts of narratives police the boundaries between nature and culture, us and them, vigilantly. And all of this works to govern how we can come to know the environment, how race and empire fit within this governing, and how we can act upon these lessons.

Rafiki's Planet Watch

Upon exiting the Pangani Forest Exploration Trail, the focus on conservation ramps up, as you are steps away from the Wildlife Express Train that takes visitors to Rafiki's Planet Watch. This "land" is named as the conservation center of Disney's Animal Kingdom (though interestingly it is not listed as a "must see" on the Animal Kingdom Web site), and serves to position Disney as a zoological and conservation expert. Hermanson suggests that this part of the park works to "invert many of the themes of fantasy and escape" (2005, 218). What Disney asserts here, then, is reality of a very particular kind: in some sense it serves to legitimate all the rest of what I have described, displaying Disney's altruistic vision for conservation. Hence, in theory the backstory disappears as guests are treated to a glimpse of Disney's work on animal research and conservation. This shift from fantasy to reality begins on the Wildlife Express, which gives visitors views of the various pens that keep the animals, framed in concrete and steel and generally inhospitable looking, all emphasizing the move from fantasy to reality, while treated to a monologue on Disney's conservation efforts. After a five-minute ride, passengers disembark and encounter a series of attractions that mete out education, Disney style. At least as symbolically important a journey as the one from the parking lot, the Wildlife Express train ride takes visitors in the heart of Disney's conservation message.

Leaving the train station behind, guests wander along a path cleverly entitled "Habitat Habit!" This trail through lush wilderness is punctuated by signs every so often, which exhort environmentalist practices, such as to "purchase wildlife-friendly products," or announce Disney's role as a conservation leader. For example, one plaque indicates, "Disney's Animal Kingdom cast is helping to save wildlife through conservation fieldwork, emergency rescues, scientific studies, and public education." If one keeps going (and there really is no choice here) you emerge into a sort of clearing, or more properly an open paved area, where visitors can look at the exotic cotton-top tamarins, and kids can learn about how they might make their backyards a habitat for local animals. Although there is slim recognition that biodiversity does occur where people live, you turn around and there is a large sign that defines conservation as "taking care of animals and the wild places where they live." Disney works again to separate nature and culture. Of course, it is a lovely idea to have backyards that are amenable to wildlife, but this is not the wildlife that Disney, or you, by virtue of visiting the Animal Kingdom, should be interested in saving. Instead, animals like the tamarins, or the creatures seen in Africa and Asia, are those in need of conservation.

All this focus on the far-distant wild is somewhat subverted by the next part of Rafiki's Planet Watch: a petting zoo called Affection Section. In this segment, kids can get up close with "exotic breeds of familiar petting zoo types: goats, sheep, pigs, etc." (Birnbaum 2005, 183). Young children were particularly captivated by this attraction, and parents waited while their kids had an opportunity to interact with rather more mundane (though still exotic) animals. No tamarins here; instead kids frolic with hyped-up versions of farmyard favorites. Sponsored by Purina, Affection Section serves to reinforce one of Disney's explicit themes at the Animal Kingdom: the love that people should naturally feel for animals, or at least for feeding them. Indeed, there is a moral proposition present in this attraction that presumes the human, however superior, must have empathy for particular, quasi-domesticated nonhumans.

The last area in Rafiki's Planet Watch, and its centerpiece, is Conservation Station: an indoor facility housing a number of different elements. As you enter the building, guests encounter a sign that flips between three statements: "a world in conflict," showing clogged highways, industrial polluters, and deforestation; "a world of choices," depicting people learning from and caring for animals; and finally "the world of animals." It is this last world that the mural that adorns the curved hallways next to this sign depicts. A remarkable piece that illustrates threatened species, the faces of pelicans, pandas, koalas, turtles, yaks, cheetahs, and gorillas stare out at the viewer, each seeming to plead for help. Pulling on the heartstrings of the guests, this mural attempts to reinforce the notion of the "world in our hands." And it works to good effect; "I want that mural in my living room," I heard one of the guests say.

As you turn the corner, you find yourself in a large room that has different areas to it. The closest section is the kiosks for the Eco-Heroes. These are phone booth–like structures where you can "speak" to a famous conservationist or biologist like Jane Goodall or rhino crusader Michael Werikhe through an interactive video display.[7] Of course, you are not actually speaking to them. Instead, it is a voice-activated recording. But it is the impression that counts (like most of Disney) rather than the actuality.

If moving in a counter-clockwise fashion, you would then encounter a series of short films narrated by Rafiki (the wise baboon from *The Lion King*) on various endangered species: wolves, koalas, bats, bears, manatees, and so on. Each of these videos shows how animals are threatened, but they also narrate success stories about some animals coming back from the brink. Very few people stay to view the entire series, but most watch one of the films and leave shaking their heads about the fate of animals in peril and the human actions that threaten them.

Next are a series of large windows where guests can observe real veterinarians at work monitoring health, administering medicines, and occasionally performing surgeries on animals.[8] On the backstage tour of the veterinary area, we were told (in a tongue in cheek way) to be on our best behavior as we too were "onstage": viewed by guests on the other side of the glass. Our guide told us that people often ask whether these veterinary procedures are "real or are part of the magic." The guide felt this attested to the good job that Disney does in creating its illusions. But what Disney does at this place is, indeed, part of the magic. It creates the illusion of Disney as scientific research institution and dedicated savior. The display of Disney's veterinary work does more than simply educate guests about animal biology. Instead, it positions Disney as a leader in animal conservation, thereby muting any questions about how it profits from captive animals.

It confirms this corporate identity with a brief display on the Disney

Wildlife Conservation Fund. This exhibit demonstrates the different areas of the world that have received Disney funds for conservation projects and encourages us to become our own eco-heroes. A short video outlines the threats to the habitat of wildlife and how Disney is working, through both research and funding, to change this.

Although Rafiki's Planet Watch is portrayed as reality rather than fantasy, it still serves as a fictive space. As Žižek remarks in *The Plague of Fantasies*, sites such as the Conservation Station, which purport to provide a behind the scenes view of the making of "the magic," actually work to strengthen the illusion: "In short, the paradox of 'the making of . . .' is the same as that of a magician who discloses the trick without dissolving the mystery of the magical effect" (1997, 102). This glimpse of the inner workings of Disney's Animal Programs is always controlled; there is no chance of seeing something untoward, for example, like the death of an animal undergoing a surgical procedure. Further, as Hermanson has also suggested, Rafiki's Planet Watch does have a backstory after all: "Conservation Station is Animal Kingdom's story about itself. Here, amidst veterinarians, computers, cameras, and petting zoos, guests are told the story of Disney's commitment to conservation and the well-being of its animal stars" (2005, 218–19).

Disney's Environmentality

I would be remiss if I neglected one of the more striking elements of Disney's work to govern and define nature and environmental problems: the remarkably titled Environmentality Program. Introduced in 1990, Environmentality, for Disney, "is a fundamental ethic that blends business growth with the conservation of natural resources" (Walt Disney Company 2009). Over the course of two decades, it has become an ensemble of texts, practices, and apparatuses that define Disney's relationship to nature. As Tom Staggs, Senior Executive Vice President and Chief Financial Officer, has noted: "Through both a wide range of entertainment products and experiences and our integrated approach to ecological issues, Disney strives to have a positive effect on our guests and consumers" (qtd. in Walt Disney Company 2005b, 4). This ecological concern is said to be predicated on Walt Disney's legacy as an environmental pioneer (Walt Disney Company 2005b.). As such, Disney's Animal Kingdom can be situated along the path laid out by Walt Disney as an expression of his love for animals.

Roy Disney notes that "he knows 'Uncle Walt' would be pleased because of Walt's life-long passion for animals which led his studio to produce more than 50 live-action animal films throughout the years" (qtd. in O'Brien 1998, 2). The Animal Kingdom also connects to the four areas where the Environmentality Program focuses its attention: wildlife conservation, resource conservation, fiscal responsibility, and environmental relations.

Animal Programs

Through the Animal Kingdom, Disney positions itself as a scientific expert in the global conservation crisis. And it begins with the appropriate credentials. The Animal Kingdom is accredited by the Association of Zoos and Aquariums (AZA), and indeed, Beth Stevens, the Vice President of the Animal Kingdom, is a past president of the AZA. The board of the Animal Kingdom is composed of "respected authorities" in conservation, including the executive directors of Conservation International and the Wildlife Society, employees of the American Museum of Natural History, academics, and staff members from various zoos (WDW News 2006d). Disney further performs its scientific expertise by asserting its leadership in research on animal species and stressing its role as a generator of conservation knowledge. Thus, we are witness to the search for new animal species during the Expedition Everest trip. Further, the Animal Kingdom has described itself as a research institution involved in the study of animal "communication, behavior, reproduction and conservation" (WDW News 2006a). Disney has a Department of Animal Programs that handles conservation education, animal behavior, animal nutrition, veterinary services, and Disney's own Wildlife Tracking Center. Moreover, Disney employs conservation biologists like Dr. Anne Savage to engage in research on tamarin "reproductive biology, social behavior, parental care and the habitat in which they live" (ibid.). These are real scientists doing real science. Through these examples, Disney asserts itself as an expert on animals. In collecting information on new species, conducting research on animal behavior, and funding other organizations that do similar work, Disney defines itself as a repository of scientific knowledge; a node for the collection of information about nature. The discovery of, research on, and display of life within the Animal Kingdom in part serve this purpose—to further knowledge about life, its characteristics, and the threats to it. By reversing their traditional reliance on fantasy and presenting "reality" to its guests, the verisimilitude of Disney's intentions regarding conservation is iterated and consumed. In connecting to the discourses of truth around science and expertise, Disney becomes not only an important agent in the production of cultural norms but also of "truths" about nature and how it can be acted upon.

But it is a particular kind of story about certain sorts of animals that is being told here. As Willis has insightfully remarked, the animals that Disney works to study and save also serve as a plot device, furthering the themed narrative of the Animal Kingdom itself: "In thematic hyperdrive, Disney's Animal Kingdom all but overwhelms the visitor with sensory cues: aroma, music, speech, dress, architecture, landscape, décor, horticulture, images, theatrics—all registers are activated to produce an unmistakably identifiable place or milieu. In this overabundant and thoroughly contrived setting, real animals cannot but be seen as elements of décor" (2005, 60). In service of the stories that Disney is interested in telling, animals must behave in nonviolent, nonsexual, easily anthropomorphized, and, above all, viewable ways. In a different context, Anderson (1995) has noted that a zoo functions not only in terms of what is visible but also through what is hidden. Thus, zoos are deeply disciplinary spaces, where the animal cannot choose its diet, when and with whom to mate, and when it will be on display or hidden. The animals are also subject to "enrichment programs," where Disney human psychologists devise ways to ensure that the animals are not bored in captivity. On a tour called Backstage Safari I was told keepers hide food in balls that the animals have to chew through, and produce bunnycicles—literally, frozen dead rabbits—that the large cats will lick all day and so remain in the view of safari riders. The animals are also given toys, like tires in the elephant pen, to keep them entertained. The guides stressed that the animals were never trained to do tricks (which omits the Flights of Wonder bird show) but rather are trained to present parts of their bodies for inspection by keepers and curators. On the same tour, one of the guides told us that in their hormone research on hippos and elephants, no animal is forced to wear the electric collars that collected information necessary for these studies—it is their own choice! How this exercise of agency takes place is unclear. But all of this is defined as enrichment.

And yet, while enrichment can be read as a strategy of dominance, I also want to suggest that it can be more than that, or rather it is a both/and kind of thing. Practices aimed at making the lives of animals in captivity better, more livable, and less, well, boring, seem to offer a possibility space for a more responsible encounter with the nonhuman. Enrichment takes note of the more ambiguous needs of nonhumans and acknowledges a reciprocal, though uneven, relationship between animal and keeper in captive facilities. It acknowledges that animals can feel, that they suffer in their habitat replicas, but that there can be pleasure and play even in captivity. Enrichment practices work to make animal lives more livable. But at the Animal Kingdom, enrichment, along with the segmentation of space, and separation of predator and prey, also means that the animals will rarely "behave badly" for the human gaze (see Wright 2006). Through their efforts at taming and managing, Disney boasts an 80 percent viewing rate, meaning that the animals are in the view of the public four-fifths of the time they are on display—a much higher average than most zoos. The guest is meant to imagine that they have seen and understood the animal as it is, rather than recomposed and reconfigured for display, part of a complex web of relationships that characterizes human/nonhuman interactions in captive facilities. At Disney, animals have been made into hyperreal replicas of themselves that are then consumed as "authentic" nature. But not only are these animals fictional, they are also highly productive, written on like a text to produce more and more and more for the pleasure of the theme park's visitors.

Of course, animals do not always behave in the ways their keepers tell them. For example, on the Wild by Design backstage tour I was told that although the "more expensive, aesthetically pleasing plants" have been electrified so that the animals cannot eat them, elephants have figured out that their tusks do not conduct electricity and manage to eat the plants anyway. On a different backstage excursion, Backstage Safari, the guests were offered the opportunity to pet a rhinoceros, who, because he was feeling ill, was "offstage." The rhino, however, had other ideas. Not interested in being touched by the visitors, the rhino, much to the dismay of his admirers, sloped off to the covered section of his pen. Then there is also the pesky native wildlife that can intrude on the natural vision that Disney presents; our guides told us that the chain-link fence with barbed wire on the top was to keep native wildlife out, because the keepers wanted to prevent them from interfering with their animal investment. And there were also animal deaths that occurred before the opening of the park: four cheetahs from drinking antifreeze, two rhinos due to medical complications, a hippo of blood poisoning, and two otters from eating too much of a particular fruit found in their now Imagineered habitat. These events, brought to the attention of the Department of Agriculture by an anonymous tip, damaged Disney's reputation as animal caretaker before the park even opened, causing a government investigation and an outcry by animal rights activists (BBC News 1998). This is wildlife untamed and unpredictable— not part of the Disney magic. All of these examples point to the slippages in the process of taming that Disney offers. And indeed, through their efforts at enrichment, it seems that at least Disney's animal keepers have a sense that their relationships with their charges are more than one way. However, it seems clear that Disney has a particular animal in mind when it shows off its efforts at wildlife conservation in the Animal Kingdom: docile, organized, and always ready to be "onstage."

Conservation

However, Disney's work on wildlife conservation does not end with the Animal Kingdom or its Animal Programs department. As the 2005 "Enviroport," the annual report on the Environmentality Program, attests: "One-third of the Walt Disney World Resort is dedicated as a conservation area. As a consequence, wildlife is abundant throughout the property" (Walt Disney Company 2005b, 23). Of course, this includes the golf courses, which are seen as sanctuaries for a range of birds and other species. Incredibly, Disney asserts: "Guests of these golf courses are learning how conservation plays a critical role in protecting wildlife habitat and natural resources. As of 2005, five of the golf courses have been certified as cooperative wildlife sanctuaries by Audubon International" (26). The dubious claim that golf courses can be seen as "natural paradises" might be tempered by other efforts. In 1993 a joint venture between the Nature Conservancy and Disney established the Disney Wilderness Preserve, 8,500 acres on the Everglades headwaters that has been protected for a range of native animal and plant species. But altruism may not have been the primary intent in the creation of the preserve. It seems as though the Disney Corporation entered into the deal so that it could expand Walt Disney World into nearby wetlands, and the Nature Conservancy made itself complicit in this project to achieve its own goals.

However, these are small examples in comparison to Disney's transformation into an eco-funding agency. The Disney Wildlife Conservation

Fund (DWCF) was established in 1995 with monies from the Disney Corporation to be a donor for NGOs working on wildlife conservation with partners in "biodiversity hotspots." It has supported the work of hundreds of people and institutions like Jane Goodall, the Dian Fossey Gorilla Fund, the World Wildlife Fund, and the National Audubon Society, doling out U.S. $10 million since its inception (when compared with revenues at U.S. $36 billion last year, a mere 0.0003 percent, Disney's commitment here does seem paltry). More recently, the DWCF gave five "conservation celebrities"—Jane Goodall, Isabella Rossellini, Wangari Maathai, Iain Douglas Hamilton, and John Cleese—U.S. $100,000 each toward their various causes (WDW News 2006b). Although the DWCF has donated a large amount of money cumulatively, the donations it gives to each organization are relatively small, somewhere in the order of less than U.S. $20,000. However, there is a range of requirements that go along with securing this rather small sum. The project must have short-term gains, must involve the community in some way, must be informed by scientific practice, and must be linked to a partner organization (Disney Wildlife Conservation Fund, n.d.). These requirements, the management of who gets money and how it is spent, are hallmarks of the kind of development governmentality that people like Escobar (1995) have documented and in which Disney wishes to participate.

Guests at Disney's Animal Kingdom and the Animal Kingdom Lodge can help to support this work through shopping. As Disney asserts: "After seeing the animals in our parks and in our 'backyard,' many of our Guests are inspired to become more involved in the effort to save our last wild places" (Disney Worldwide Outreach, n.d.). By involvement, however, Disney means donating one dollar to the DWCF when you are making another purchase at select parts of the Disney universe, and then proudly displaying a trading pin, which proclaims you a "conservation hero." Thus, guests can feel good about emptying their wallets to purchase all manner of Disney paraphernalia, as it is done "for a good cause." Guests who add a dollar to their purchases are able to assume, at least within the various Disney parks, the guise of an environmentalist while reinforcing the same consumptive patterns that have in large part led to the environmental crisis that we now face. Through "lands" like Rafiki's Planet Watch, the research they conduct on animals, and the DWCF, Disney combines care for animals, science, education, and the market to make the "truth" about nature and govern the way this "truth" is understood.

Resource Conservation and Fiscal Responsibility

Two other aspects of Disney's Environmentality Program seek to reinforce these notions and position the corporation as a key actor in defining and working on environmental problems: resource conservation and fiscal responsibility. I group them together because for Disney they seem inextricably linked. Disney has turned to green purchasing, has reduced their energy expenditure by 3 percent, has started composting, implemented reusable coffee mugs, and makes sure copying is done on both sides (Malmberg 1998; Walt Disney Company 2005b). It is not my intention to sound dismissive of these efforts. Indeed, I would strongly suggest that these are not bad initiatives; green purchasing, reduction in energy consumption, and recycling should be adopted as tools to limit ecological footprints, especially those of corporations. And Disney has shown that these efforts at resource conservation result in real monetary gains; in 2004, because of these projects, Disney saved more than U.S. $38 million, which, you will recall, is more than it has given away through the DWCF (Walt Disney Company 2005b, 5). The "2005 Enviroport" argues: "Disney continues to demonstrate that the use of technology and creativity within the global Environmentality arena can deliver consistent financial successes" (5). The stress on technology utilizes the discourse of sustainable development that has so infused the mainstream environmental movement. Further, Disney's emphasis on the benefits of restraint reinforces the notion that practices of consumption should be the main framework through which we act upon environmental problems.

Environmental Relations

The last element of Disney's program is what it terms environmental relations, meaning the host of programs, internal and external, that promote green-mindedness. So, Disney has a whole range of educational and outreach activities that encourage the development of environmental subjectivities. Disney has established the Jiminy Cricket Environmentality Challenge, in which fifth graders vie for the grand prize by taking an "Environmentality pledge" and develop a plan to address an environmental issue of their choosing. The prize? A trip to Walt Disney World or Disneyland, of course! Disney has also established camping programs for urban youth in their own communities to teach marginalized kids about nature. This program now operates in Florida, California, Hong Kong, and the Cayman Islands. Disney further encourages guests and cast members to participate in the Environmentality Program. For staff, there is a two-day training program on green practice that culminates in the signing of "an *Environmentality* Pledge that commits them to maintain an awareness of the wise use of resources" (Malmberg 1998, 164). For guests at the Animal Kingdom, the curator of education for guest services trains Disney employees to "strike up 'conservation conversations'" at various sites in the park (164). And for "cast members," guests, and the general public alike, "Disney's Commitment to Conservation" (Walt Disney Company 2005a), a glossy report outlining Disney's work on environmental issues, provides some easy tips for wildlife conservation action. Much like the American Museum, Disney provides manageable steps such as subscribing to wildlife magazines, encouraging recycling and reusing, avoiding exotic pet species, purchasing green products, and having a bake sale for local environmental organizations.

What Disney works to govern here is the formation of a particular subject–position: the green consumer. So what? Isn't it better to have people who are conscious of environmental issues? The problem is that Disney's environmental subjectivity stays quite securely within the boundaries of mainstream critique and market-based capitalism and, in doing so, is able to offer a position of relative comfort to those who wish to be environmentally minded. You do not really have to do much to save the earth. Throw a bake sale, recycle your cans, give money to a nongovernmental organization. Not mentioned are the environmental costs of suburban life, the American reliance on oil, or the complicated impacts of resource extraction, and the fact that one has to pay eighty-seven dollars for the green lesson. Instead, there is an easy, quick fix that does nothing to take apart the neoliberal order that both creates and increasingly commodifies environmental problems. Disney structures vision and understanding so that nature, and its rescue from "us," becomes experienced through consumption.

The company goes to great lengths to ensure that the park is read in this way. While saying this, it is not my intention to suggest that there can be no alternate readings. My very presence in the park asserts that there can. Of course, people can and do subvert the primary message that the Animal Kingdom, or perhaps the Environmentality Program more generally, produces. So, for example, teenagers look suitably bored with the whole affair, stopping to laugh at the costumes Disney has dressed its "cast members" in. Others with a more critical lens might understand the eco-tourism theme to be a thinly veiled form of imperialism. What I am not doing here is suggesting that people are dupes to corporate marketing campaigns. Rather, what I want to submit is that Disney arranges its narratives in such a way as to make them compelling, and they attempt to control the way people experience the stories, images, experiences, sensations, and feelings.

Conclusion

Disney's Animal Kingdom is a complicated and contradictory place, blurring the boundaries between real and hyperreal, artificial and natural, commodity and affect, capitalism and conservationism, romanticism and biology. But this does not mean that Disney's nature narrative lacks coherence. Instead, drawing on all these registers is what makes Disney one of the most important brands in tourism. And it is because of Disney's influence that it becomes an important agent in green governmentality. As Michael Hutchins, advisory board member to the Animal Kingdom and AZA member, states: "What can Disney do for conservation? Disney can make it a household word" (qtd. in Malmberg 1998, 33). So, one cannot simply write Disney's Animal Kingdom off as just a space of mindless entertainment: it is a site where "truths" about nature are produced and circulated.

Stephen Fjellman, in his oft-cited treatise on Walt Disney World called *Vinyl Leaves*, expresses his view that "with its various theme parks and hotel complexes, [Walt Disney World] is the most ideologically important piece of land in the United States. What goes on here is the quintessence of the American way" (1992, 10). While I think the concept of ideological space is useful, I fear that Fjellman's assertion remains essentially unprovable. What of Washington, D.C., the Pentagon, Harvard University, Wall Street, Madison Avenue, Hollywood, Yellowstone National Park, the U.S.-Mexico border, and the other sites in the United States that carry the freight of ideological production, economic and political import, as well as cultural meaning? How and in what context can we define what is *the* most ideo-logically important? I think we need to leave this kind of hierarchical ranking aside in favor of the notion that knowledge and power are produced and circulate from multiple sites. The real question, then, is what kind of space are we dealing with in this particular context; what makes Disney's Animal Kingdom different not only from other kinds of productive spaces but also from the rest of Walt Disney World? I suggest the answer is that the Animal Kingdom offers an excellent example of a corporate biopolitical institution, and as such is a central signifier of what the governmentality of nature means.

Paul Rabinow and Nikolas Rose, in an article entitled "Thoughts on the Concept of Biopower Today" (2003), mark out the territory they feel is necessary to an exploration of biopolitical projects. They argue that an analysis of biopower must include discourses of truth exercised by specific experts, strategies of intervention on target populations, and specific kinds of subject formation. Usually, these indicators are applied to political projects rather than corporate endeavors. But by way of conclusion, I use Rabinow and Rose's insightful heuristic to underline the assertions I have made about Disney's ability to govern the understanding and consumption of nature, extending how we might conceive of the biopolitical realm.

Throughout the course of this chapter, I have emphasized the importance of Disney's reliance on experts and specialists to justify and explain the realm of the Animal Kingdom. As we have seen, various biologists, conservationists, zoo operators, and academics were involved in the development of its broader vision and now the day-to-day operations of this theme park. Disney hosts an Animal Programs Department that undertakes research in different aspects of animal life. It embarks on scientific expeditions to "discover" new species of plants and animals. Disney makes its veterinary operations visible through Conservation Station, literally performing their scientific credibility. The reliance on these real/imaginary experts in the real/imagined field of conservation works to position Disney as a soothsayer in the game of speaking for nature. By utilizing discourses like scientific expertise, Disney is able to legitimize its practice and contribute to truth-making about nature and conservation. In doing so, Disney makes loud claims of "authenticity," the kind that are not foreign to the scientific endeavor. Disney's Animal Kingdom presumes and legitimates an objective set of authoritative master witnesses who are able to see, understand, replicate, and represent what animal reality actually is for the consumption of their guests.

But what do these authorities produce? Of course, as I have argued all along, they participate in an already existing discourse of truth: that of scientific expertise and objectivity. And yet, this objectivity is complicated at the Animal Kingdom by the thing that Disney cannot do without—fantasy. Indeed, at times this site employs a discourse of "un-truth." Fiction and fact are confused in the realm of the Disney universe; even the real is fictionalized, and vice versa. So, on one hand, guests see displays of veterinary skill, while on the other they embark on a journey back through time to rescue a dinosaur and vindicate the scientific endeavor. A reproduction of Mount Everest replete with raging yeti is juxtaposed with Disney's actual "expedition." A real African safari is rendered unrealistic in comparison to its facsimile. Guests are meant to derive conservation messages from the theatrical enactments of *The Lion King* and *Pocahontas*. The Tree of Life, perhaps the penultimate expression of fantasy, stands as an emblem of ecological interconnection. All this and more—the work of the Imagineers, the Disney characters, the tie-in to Disney fiction—demonstrates how fantasy can act as a vehicle of "truth" because it shows guests both how the world currently is, and how it should or could be. Put another way, this discourse is not only ideological, referring back to Fjellman, but it is dystopian and utopian, drawing upon anxieties and hopes to tell a story about the current state of nature, and how it could be, if only people realized that Disney offers solutions to the environmental ills of the world. What makes the Animal Kingdom distinct among the panoply of biopolitical institutions, then, is its use of fantasy—and through fantasy an appeal to play and pleasure—to tell the "truth" of nature.

Rabinow and Rose further suggest that biopolitical projects must involve particular "strategies of intervention upon collective existence in the name of life and health" (2003, 2) that can occur across a variety of scales and in a myriad of spaces. What makes Disney's Animal Kingdom so interesting as an agent of biopower is that its strategies of intervention target two different yet interrelated populations: the human and the nonhuman.

For the human visitors to the park, the Animal Kingdom's most potent strategy is the evocation of affect. The Animal Kingdom uses spectacle to exploit and condition what are deemed to be significant aspects of people's emotional and affective condition. For example, the Tree of Life is meant to elicit wonder and awe in the miraculous works of nature. The petting zoo in Rafiki's Planet Watch serves to induce love and empathy in the small guests who partake of its pleasures. The whole of Rafiki's Planet Watch works through curiosity: the desire to know about the earth. The story of Big Red and Little Red, the two elephants at the center of the Kilimanjaro Safari ride, works to arouse both a sense of anxiety at the possibility of their slaughter and a feeling of hope once the guests have participated in their rescue. Similarly, the Kali River Rapids ride functions as an engine of nostalgia (perhaps for the colonial contest), anxiety (about global deforestation), and hope and pride (that ecotourism can stem the tide of ecological devastation). Each one of these cases works through some sort of spectacle to evoke emotions, which have an impact on how people come to see nature and conservation. These spectacles are backed up by narrative, which attempts to predetermine understanding, to establish scripts, to offer conceptual categories and fix terminology, indeed to name and classify, all of which are crucial to the workings of biopolitics.

While the Animal Kingdom attempts to govern those who enter its gates, it simultaneously works to make up those who remain outside of them through the act of consumption. As mentioned previously, Disney positions its guests to consume racialized others as hyperreal commodities: Pocahontas as Indigenous environmental princess; Asians as temple-going ecotour operators; Africans as either poachers or safari guides. In each of these cases, those represented are not the authors of their own bodies (to paraphrase Donna Haraway 1989) but rather are made through the images, imaginings, and fantasies of the white bourgeois public that has been assembled to consume them via the gaze. Thus, Disney's governing stretches beyond its physical boundaries to circumscribe and delimit through representation the ways that people who will never set foot on a Disney property are understood and experienced.

The human population to which Disney applies its technologies of power is presumed to be active—that is, they have the capacity to do. The other population that Disney targets is generally not afforded the same possibility. Animals in the theme park are seen as essentially passive; objects that can evoke action in humans but must be subject to disciplinary measures. In the strategies it employs to manage its animal populations, Disney enacts its own rituals of purification in the Latourian (1993) sense of the word; the division between culture and nature here is strictly policed. Of course, we also know from Latour that this kind of purification can only lead to the proliferation of hybrids, or what he calls, in his rather unwieldy way, quasi-objects. However, the Animal Kingdom acts as Latour's moderns do: relying on the entanglements of nature and culture while at the same time denying their existence. The animals are thus a product of the imbrications of nature and culture; they are meant to be read as purely natural entities but in fact have been thoroughly remade not only for the pleasures of human vision, but also to teach humans about their own culture. Thus, the animals at this site can only speak through Disney and its associates like the Disney Wildlife Conservation Fund or Conservation International. So we certainly have a population here: one that is territorialized and classified, named and even numbered, and granted a biosocial status, but one that is controlled, that is manipulated, that is represented as docile and entertaining in ways that wild nature can never be, but one that is simultaneously not completely dominated.

Finally, Rabinow and Rose (2003) point to the importance of "modes of subjectification" to the project of biopolitics. They note that these are the means through "which individuals can be brought to work on themselves, under certain forms of authority, in relation to truth discourses, by means of practices of the self, in the name of individual or collective life or health" (2003, 3). What should be clear by now is that the subject–position proffered by Disney is that of the green citizen. However, it is a particular kind of citizenship where consumption becomes the rite of belonging. Indeed, through the purchase of park admission a potential green citizen is born. But more than this, Disney's Animal Kingdom provides a list of techniques that individuals can employ to become better environmentalists: contributing to the Disney Wildlife Conservation Fund and other conservation organizations, purchasing green products, engaging in Environmentality challenges, making our backyards welcoming to animals. These are the technologies of the self that Disney affords its guests to achieve actualization as a green subject, all of which preclude collective action.

Of course, there are some limits to who can claim this identity. As explored in the first section of this chapter, Disney's Animal Kingdom is a white bourgeois space. The green citizen Disney helps to define is also very much an affluent one who is expected to use her or his means to aid nature and the animals. So, for example, the kind of environmental justice work that pays attention to the intersections of race, class, gender, and nature is not present. Nor is a sustained critique of industrial capitalism and its effects on nature. A philosophy of animal rights is missing. In short, the multiplicity of subject–positions offered under the umbrella term "environmentalist" are occluded in favor of the Disneyfied version of what it means to care about nature.

However compelling, the intended effects of these technologies may also be somewhat fleeting. In a 2001 report published by *The Informal Learning Review,* members of Walt Disney World Animal Programs document a study they conducted, interviewing 501 guests, 30.5 percent of whom were contacted two to three months later to comment on the effect of Conservation Station on their sense of environmentalism. It appears that the visitors to this area of the park are a self-selecting bunch, who are "well-educated (83% had at least some college); almost half belong to nature or conservation organizations (46%); and over half visited a zoo or aquarium in the last year (64%)" (Ogden et al. 2001, 18). In some sense, many visitors had a baseline understanding of the kind of conservation message that Disney espouses. Although guests understood and agreed with what is offered as the main message of the site—the importance of conservation—in most cases, this understanding did not translate into lasting change. For example, the report notes that three months later, the desire to discuss conservation with others had waned (18). So it seems as though the technologies of the self proposed by Disney as the answer to nature's ills are not always taken up once the visitors have reentered their everyday space.

One wonders, however, if the story can be made a bit more complicated. I have argued that the message at Disney's Animal Kingdom is not solely about environmentalism, but also about consumption. Given this assertion, it would be interesting to ask guests how many of them have given or continue to give money to conservation organizations rather than cleaning up a wetland, or how many of them have purchased tickets to an ecotour rather than engaging in a discussion of animal rights. As Hermanson has caustically remarked, "The contradictions that would seemingly tear apart the fantasy of the Animal Kingdom—ludicrousness of a hyperconsumerist theme park spouting ecological messages—are no more troublesome for the typical Disney tourist than the contradictions of the Sierra Club member driving an SUV" (2005, 201). The best, most productive technology that Disney offers, I'd wager, is the ability to buy. Perhaps after paying eighty-seven dollars per person, the consumer has bought the right to forget all about conservation. The easy answers to the anxiety around environmental devastation are the Animal Kingdom's lasting contribution.

What is significant here is that these discourses, strategies, and modes of subjectification demonstrate how a nonstate authority, in this case a corporate entity, can so successfully organize a biopolitical project, and because of this, function as an agent of green governmentality, all the while generating profit by deploying nature and eliciting emotional reactions to it. Disney has constructed out of science and fantasy a new kind of environment that has proven exceptionally appealing, although its effects are short term. In its appeal, power is exercised on individuals and populations to tell the truth of a particular kind of nature where spectacle and education is repackaged as fun. Disney's Animal Kingdom stands as a materialization of the effort to exercise power over life in order to produce profit, pleasure, and particular people. In doing so, the Animal Kingdom serves as dramatic evidence of the amazing capacity of capitalism to innovate and colonize, to expand in this case into nature, which is perhaps something of a biopolitical frontier.

As Susan Willis has noted, "At Disney, conservation is more than narrative; it's a spectacle" (2005, 56). Animals on display and performing for crowds, an enormous fake tree and simulated safaris, yetis and dinosaurs, observing veterinarians at work and watching mole rats behind glass at "research stations": all these elements act in service of this notion of conservation as spectacle. But Disney's work at the Animal Kingdom goes beyond spectacle to govern those who consume it. The larger point is that Disney commodifies our vision and understanding, and through that commodification, governs what we come to know as nature. As Hermanson (2005), Price (1995), Cronon (1996), Davis (1997), and others have reflected in different contexts, this works in large part because nature has already become a theme park, or perhaps a mausoleum. Nature is where people are not. It is the journey to find nature at a zoo, a theme park, an ecotour, or a museum that has allowed for it to be consumed in normalized ways by certain people. In her work on Sea World, Susan Davis (1997) has insightfully remarked that what needs to be queried about places like Disney's Animal Kingdom is what stories are occluded in this theming of nature and how this erases possibilities for it to be otherwise. The next chapter picks up on these leitmotifs.

CHAPTER 3

Wolves, Bison, and Bears, Oh My!
Defining Nature at Yellowstone and Grand Teton National Parks

*Thousands of tired, nerve-shaken, over-civilized people are begin-
ning to find out that going to the mountain is going home; that wildness is a necessity; that mountain parks and reservations are useful
not only as fountains of timber and irrigating rivers, but as fountains
of life.*

—John Muir, 1898

IN OCTOBER 2006, I embarked on an ecotour to Yellowstone and Grand
Teton National Parks in Wyoming: a seven-day excursion into "yester-
year's frontier." On the last day, I sat in the van that had transported our
ecotour all over the two national parks, pondering our trip. I gazed out the
window, half listening to our guide explain the impact of brucellosis, an
infectious disease that induces spontaneous abortion in both livestock and
"wild" creatures like bison. It has been a remarkable trip, I thought, filled
with spectacular mountain vistas, quaint towns, and a picnic in the snow,
as well as sightings of bears, wolves, bison, elk, eagles, and moose. Postcard
perfect. All the other tour members seemed thrilled with their experience,
with some "best trip ever" comments; no small praise given that most had
gone on ecotours throughout the globe. Why not? It had lived up to the
claims in the brochure. But I wondered, just what was "wild" and "natural"
about this trip? Is a national park natural? How natural is it for a herd of
bison to cause a traffic jam or a herd of "wild" ecotourists to take pictures
of it? Is the wildlife any less incarcerated than the ones seen at Disney's
version of a zoo? How do pleasure, play, and adventure function in this
space? How does tourist entitlement work here? Does this experience
limit what can be understood as nature? And if it does, through what technologies and practices is a particular kind of green governmentality made?
And more practically, just how many toxins was the van that we toured

around in spewing out so that we could have this opportunity to gaze upon nature?

This chapter addresses the above questions to explore how ecotourism is not only about environmentalism but also about power; or, more properly, that it is sometimes difficult to separate the two. I argue that ecotours can work to structure vision, separate nature and culture, and proffer discourses of science and romanticism that have the effect of "truth." In this way, ecotourism, like the other case studies of this book, seeks to govern what we understand and experience as nature, leaving little room for nature to be elsewhere or otherwise. And it appears that ecotourism, although once touted as the solution to the negative impacts of mass tourism, can be as much about capitalism and consuming the "other" as is a resort or a cruise holiday.

But I also want to say something a little more specific about how green governmentality is manifested and circulates through the practice of touring the wild. Unlike the other cases in this book, I suggest that this case study offers up an opportunity to think with the notion that green governmentality can work on an aesthetic register, not only governing the production of knowledge but also how we come to see the wilderness as beautiful and rejuvenating, and because of these characteristics, in need of human protection. As such, I contend that this ecotour offers a kind of visual grammar for natural beauty, backed up by a long history of romantic encounters with wilderness and narratives of a national nature. Working through an assemblage of lens/eye/animal, the ecotour is a veritable school for producing subjects to learn the appropriate way to see nature and reproduce it, ensuring that such vision travels beyond the boundaries of the national parks.

But how does this seeing work? Here I draw on studies critical of the tourist endeavor to bring ecotourism in conversation with green governmentality. John Urry's (2002) "tourist gaze" is a central concept in my analysis, not only because some of its insights are generated from Foucault, but also because it emphasizes visual consumption as central to the tourist experience. Foucault argued in *The Birth of the Clinic* (2003b) that an objective, observing, professionalized, and improving medical gaze emerged in the eighteenth century, which brought the body into relations of power and discourse in new ways. Although Urry notes that one would imagine sightseeing to be far removed from the sterile environs of a clinic, he suggests that the gaze of the tourist "is as socially organized and systematized as is the gaze of the medic" (2002, 1). While there can only be a wide diversity in the ways that people act as tourists, Urry (2002) contends that the tourist gaze can be understood within a framework of attributes that define modern leisure travel. First, Urry asserts that tourism is that which "presupposes its opposite, namely regulated and organized work" (2). It is a leisure activity that is removed from the ordinary, everyday life of labor. This separation involves movement for Urry, to places and spaces that are different from the ones we call home. But this movement is by definition temporary and necessitates a return to the ordinary. Thus, the break from everyday life is key to tourism. It is leisure rather than toil. And it is because of this time of respite that the tourist gaze is a lingering one, examining each element of the touristscape that has been deemed worthy of a visit. This prolonged gaze is searching for the "it-ness" of a place; the characteristics that can be distilled to a particular set of signs. So, tourists seek essences that define places in their imagination: romantic, rustic, wild, sophisticated, bucolic, urban, and so on. People then collect and memorialize these sites through visual media like photography that "enable the gaze to be endlessly reproduced and recaptured" (3). And all of this is undergirded by the tourist industry that makes and remakes places to develop new tastes, trends, and above all, to direct the tourist gaze. A visit to sites deemed important by the tourist industry—of spectacular nature, high culture, ancient history, or even extreme poverty—works as a trophy, an iconic place to check off on a list of must see places. We might situate ecotourism as part of this project of remaking, both as a new trend in tourism and a different way to gaze upon places. This way of seeing and reading particular places can be a fruitful way to talk about the governance of nature as well as its commodification because the acceleration of consumable signs may actually lead to a more limited number of meanings.

However, my use of Urry's tourist gaze should not be understood as an acceptance of a monolithic understanding of the tourist, or his or her activities. Indeed, the signs consumed are never stable and are often jettisoned for new meanings altogether. So, where once Las Vegas was seen as a seedy home to forlorn gamblers, it now has been recuperated as a sophisticated while still risqué destination, playing on its former reputation. Further, tourists themselves do not passively encounter the sights they seek out, but rather are often playful, ironic, and fully aware of the unreality of what they consume. If not, places like Disney's Animal Kingdom would receive very few visitors. Tourism is an encounter rather than a vehicle of transmission, and both site and subject are changed in the process. Nonetheless, amid these contingent meanings and wry readings, there is an objectifying gaze through which tourist sites are *meant* to be understood. This gaze can operate to fix dominant meanings of a place, indeed works to produce the "reality" through which a space can be encountered.

Tourism has also been described as a thoroughly modern activity (Desmond 1999; MacCannell 1989; Sandilands 2003; Urry 2002). The modern nature of tourism is predicated by a sense of nostalgia, mourning, or loss, which scholars like Braun (2002) have shown is central to the experience of the ecotourist. Braun argues that the modern subject, having transcended the past and always seeing the present as an eclipse of the past, experiences modernity through loss. In this articulation, pleasure is found when the ecotraveler searches for pristine beginnings—for nature untouched by modernity—which, in fact, can never be found. Thus, Braun argues, "adventure travel and ecotourism are best viewed as practices through which subjects perform and reaffirm the present—and their own identity—as modern" (2002, 112). This kind of travel does not step outside of modernity, as seems to be the purpose in attempting to encounter an "authentic" and primeval nature, but rather is an expression of it. Along with the sense of loss, modernity comes with a whole host of other elements that inform the tourist endeavor: the co-emergence of photography as a way of experiencing tourist sites; the sense of entitlement to come upon the "other" that mass tourism has produced; the emergence of striking boundaries between what is considered work and play, labor and leisure, everyday life and spectacular experience; and, of course, tourism's insertion into global capitalist economies.

Scholars have done an admirable job in focusing on nature's commodification through tourism (Duffy 2002; Fennell 1999; Honey 1999; West and Carrier 2004), naming and exploring how ecotourism has inserted nature into global capitalism in novel forms. These studies are extremely useful because they have questioned the supposed innocence of the ecotourist endeavor, questioning their claims of environmental stewardship and sustainability. My goal is to push this interrogation further. I seek to not only discuss ecotourist nature as a commodity but also as a discursive site, which can construct "truth," name risk, and proscribe behavior. So, my point here is not to say that ecotourism isn't about capitalism, but it is also about more than money: it is about the way we encounter and understand nature because of the narrative work of these enterprises. Thus ecotours, like other vehicles of recreational culture, are more than what they appear at first blush. The selling of nature as commodity, combined with a moralizing environmentalism that seeks to obscure its commodification, shapes ways of seeing nature.

And so, I am interested in using these insights to think about how tourism and ecotravel operate in the American context. However, I also want to complicate these notions by considering how these dynamics work in a site of national nature. The pilgrimage to visit national parks is a particular one, awash with notions of nationalism, wilderness, and its mastering through colonialism. Moreover, western parks like Yellowstone and Grand Teton are host to notions of manifest destiny, the frontier, "cowboys and Indians," and the discursive as well as political struggles over charismatic megafauna like bison, wolves, elk, and bears. An examination of ecotourism in a national park, then, allows for the considerations of how these "sacred" sites in the United States are defined by the complex entanglements of nature, history, politics, commerce, and power.

Ecotourism

Mass tourism has expanded to proportions that were likely unimaginable at the beginning of the twentieth century. Estimates indicate that approximately 880 million people traveled internationally in 2009 (UNWTO 2010). Tourism has been named the world's largest employer, involving 10 percent or 200 million jobs worldwide (International Ecotourism Society 2000). The tourist industry has grown to such a degree that theorists like Urry suggest that "[t]o be a tourist, to look on landscapes with interest and curiosity (and then to be provided with many other related services), has become a right of citizenship from which few in the 'west' are formally excluded" (1992, 4).

Ecotourism is, of course, a subset of the larger tourist industry, and as a concept and practice it has generated much rancorous debate in recent years, both in terms of how it can be defined and what it purports to do. Probably the most widely accepted definition of the term comes from The International Ecotourism Society (TIES). According to this network of tour operators, conservation organizations, nongovernmental organization staff, government workers, academics, and travel agents, ecotourism is "responsible travel to natural areas which conserves the environment and sustains the well-being of local people" (International Ecotourism Society

2000)—a broad definition that in the end says very little about actual practice. Perhaps because this definition is so fluid—it can include all manner
of outdoor recreation, adventure and extreme travel, visits to national
parks, eco-volunteering, and wildlife watching—the statistics around ecotourism's growth seem so variable and high. So, for example, the World
Resources Institute pegged tourism's growth at 7 percent in 1990, and
ecotravel was said to be expanding somewhere in the order of 10 to 30 percent. In the United States, the market for ecotourism is estimated at $77
billion per year, some 5 percent of the total U.S. tourism market (International Ecotourism Society 2000). Within this segment, national parks
figure large. In 2004 national parks received 277 million visitors, and are
said to have "generated *direct* and *indirect* economic impact for local communities of US $14.2 billion and supported almost 300,000 tourist-related
jobs during 1996" (ibid., emphasis added). While the amounts garnered
from ecotourists, or what is meant by direct and indirect, are unclear, sites
deemed natural appear to be drawing more and more visitors per year.

More broadly, despite the differences in definitions and their attendant
impacts on how the scale of ecotourism can be measured, what cannot be
disputed is that the ecotourism industry has seen massive gains and generated much interest, especially among those who see themselves as socially
and/or environmentally minded. In a discussion of the rise of luxury eco-
hotels, Heidi Mitchell (2006) argues in the *New York Times Magazine* that
being green is becoming the "new normal" in an attempt to capture the
"metrospiritual" market share, quite clearly a play on Mark Simpson's 1990s
neologism "metrosexual": a "feminized" male subject (gay, straight, bisex-
ual, or otherwise) who rather than performing the strict division between
gender roles around normative conduct, instead embraces self-care, style,
and emotion. According to Simpson, metrosexuality emerged in the 1980s
and 1990s, when "old-fashioned (re)productive, repressed, unmoisturized
heterosexuality [was] … given the pink slip by consumer capitalism" (2002).
Instead, a whole range of consumer products emerged to help men to
achieve aesthetic perfection, from personal training to "manscaping." Indeed, Flocker's book, *The Metrosexual Guide to Style* (2003), contains chapters on etiquette, fine food and wine, art and culture, grooming, fitness,
and romance, all categories in which the metrosexual must excel. Naming it
the beginning of a new consumer category, Simpson remarks that "Metrosexual man is a commodity fetishist: a collector of fantasies about the male
sold to him by advertising" (2006). Thus, the metrosexual is a complex

figure, both disrupting gender and sexuality boundaries, all the while becoming more deeply intertwined with capitalist markets. If these characteristics are ascribed to the metrosexual, who, then, are these so-called "metrospirituals"? According to Mitchell they are "hybrid-driving, yoga-practicing baby- and echo-boomers for whom social responsibility and seeking out new adventures are a way of life" (2006, 14). Metrospirtuals eat local foods, support green organizations, and consume responsibly. Similar to their namesakes, metrospirituals have a sensitive side, seeking both communion with nature and its preservation. And like metrosexuals, they too are a consumer category that is targeted by very particular marketing strategies, specifically around style, sensory experience, and adventure, and they use their purchasing power to express their vision of the world.

Besides the characteristics named above, typical ecotourists tend to be between thirty-five and fifty-four, college educated (82 percent),[1] split evenly between men and women (depending upon the activity in which they are engaged), and more likely to be willing to spend up to $1,500 more on such excursions (International Ecotourism Society 2006). The desire for social and environmental responsibility underpins the marketing of these adventures, and the people who go on them often seem genuinely interested in traveling ethically.

But this kind of ethics—the sense that power relations can be shifted, ameliorated, or overcome through "political consumerism"—is the core of ecotourism that receives the most criticism. The practice has been assailed in recent years for producing exactly the opposite of what it purports; rather than improving conditions for local people and safeguarding the environment, ecotourism can lead to an increase in contingent, low-paid labor and environmental destruction (Honey 1999). More theoretically, ecotourism is said to espouse a kind of "weak sustainability," which, in fact, reinforces the commodification of nature (Duffy 2002, 155; West and Carrier 2004). Indeed, what I find most useful to my analysis of ecotourism in the United States is the connection made between neoliberal capitalism and the remaking of nature for consumption. It seems in some ways that ecotourism is an almost inevitable function of the colonizing spirit of capital. Environmentalism and marketing, nature and profit, together at last—as if they were ever apart.

Catriona Sandilands's (2003) work on the reimagining of Clayoquot Sound's landscape and economy has been particularly instructive, situating ecotourism squarely within global capitalism. In her examination of the shift from extractive to attractive, tourist economies, Sandilands charts how preservation can operate through the same logic as clear-cutting: both serve to weave this piece of temperate rainforest into capitalist relations, albeit with dramatically different outcomes. But the trick here is that it seems not to be so. As Sandilands contends, the maintenance of a landscape ripe for the tourist gaze is made innocent of these economic considerations: "The problem is rather that this representation comes disguised as a liberation: one set of capitalist-embedded (consumer) constructions of nature gets to pass as a freeing of the landscape where another, less romantic (productive) aesthetic is demonized as if it were the only representative of multinational capitalism around" (2003, 141). Through this movement from productive to preserved, places like Clayoquot Sound become remarkable, remade in the image of the global commodity of nature: pristine, primeval, wild. In doing so, this site defined as nature must first be emptied of cultural or economic traces. Sandilands advances the notion that what makes such wilderness spaces a tourist destination is the fact that they are divorced from everyday life, removed from work, home, and livelihood. What Sandilands describes is a place that, in some sense, has been made into a museum.

Nature's Nation: U.S. National Parks

Perhaps one of the most memorialized spaces in America is the national park. So-called "white settler" nations like the United States expanded and flourished through the imperial project of conquering and taming a supposed wilderness. But wilderness has not always carried the same freight in American imaginations. William Cronon (1996), in his now famous essay "The Trouble with Wilderness," charts the discursive shift about notions of nature that was fully realized in the United States at the end of the nineteenth century. Previously, wild places were seen as a wasteland. Ideas about wilderness were drawn from the Bible, filled with temptation, dread, and "where it was all too easy to lose oneself in moral confusion and despair" (70). However, Cronon argues that wilderness was recuperated through the interrelated social constructs of the sublime and the frontier. Drawing on the same biblical imagery but inverting its meaning, wilderness became not only a space of encounter with evil but also with the divine. Put more succinctly, "If Satan was there, then so was Christ" (73). Thus, the wilds of nature became a place where one could encounter God.

But these were more than just religious landscapes; they were also national ones, and part of the making of the nation in the United States involved the myth of the frontier, and the lament for its loss. As Cronon shows, the frontier has special resonance in the iconography of the United States, linked as it was to nation building, masculinity, and rugged individualism. However, eventually modernity infected the West and the frontier "way of life" faded. The attempt to recapture an echo of the frontier, a highly productive discursive and material space that had been so central to the emergence of an American identity (or at least so Frederick Jackson Turner [1986] argued) was something that could only take place through wilderness. This different vision of nature—pristine and rejuvenating rather than wicked and villainous—is what in large part served as the impetus to establish the first national parks, beginning with Yellowstone in 1872. They acted as preserved pieces of a lost frontier, incarcerated in space and time as symbols of U.S. identity.

The opportunity to visit the canyons, trees, mountains, and waterfalls of the first national parks was initially limited to a select few: those of the leisure class who could afford transcontinental travel. However, Marguerite Schaffer, in her book *See America First* (2001), charts the ways that national parks were opened up to a broad audience. Beginning in the 1880s, national parks were linked to the burgeoning westward rail system, offering a means to travel to these natural wonders. With the founding of the National Park Service (NPS) in 1916, a new player joined the game of making these natures iconic. In the 1920s the NPS began a campaign to encourage Americans to see their country first, to visit and understand its natural features as a ritual of citizenship that attested to the greatness of the nation. Both the railway barons and the NPS employed massive advertising efforts to lure people west to good effect. Schaffer asserts that the infrastructure of the modern nation-state—railways, roads, telephones—made these natural wonders accessible to middle-class white Americans, and generated the notion of a national tourism that centered on the parks. This connection between nationalism, nature, and citizenship cemented in places like Yellowstone is reflected in the oft-quoted remark by western novelist Wallace Stegner, who opined, "National parks are the best idea we ever had. Absolutely American, absolutely democratic, they reflect us at our best rather than our worst" (1983, 4–5). Putting aside the many ways one might take apart this statement, Stegner accesses one main function of national parks:

to define a sense of "Americanness" and the West through the consumption of this simulacral frontier.

If national parks have been host to notions of nature and nation, they have also been witness to how these concepts have been worked through performances of whiteness, masculinity, sexuality, and class. Here, the perfect figure for such assertions is Theodore Roosevelt. From a prominent New York family, Roosevelt was raised within a bubble of privilege and power, and from him great things were expected. However, Roosevelt suffered terribly from asthma beginning in his early childhood into his adult years. This ailment marked him as an invalid who accompanied his mother, Mittie, on her jaunts to resort spas to recover rather than engaging in the manly pursuits his father so venerated (Dalton 2003). But rather than understanding the physical basis for his affliction, his doctor diagnosed Roosevelt with "the handicap of riches" brought on by "excessive upper-class refinement" (37). Named a sissy, sickly, and weak, Roosevelt could not claim the masculinity that he so craved. The notion that disease was related to the "dandyfication" of the upper classes was in turn linked to other problems with the social body—panic about racial decline in the face of waves of immigration, anxiety around the effeminacy and over-civilization and decadence of the city, and fear about the education of women, which might subvert their reproductive functions—all which led to a kind of "national emasculation" (39; see also Haraway 1989). According to Dalton as well as other biographers, this assessment sharply marked Roosevelt's perception of both his illness and the state of the nation. But there was an antidote to the anxieties of the nation: masculine pursuit of nature's mastery, particularly in frontier-like wilderness of the West. A "strenuous life" was the balm needed in this quest to recover manliness, class power, whiteness, and sexual potency. And so, Roosevelt set out to encounter wilderness, first in the Dakotas for ranching and hunting, but later in places like Yellowstone, Yosemite, and East Africa.

However, these practices of self- and nation-building could only happen if these sites of nature were preserved. And so, Roosevelt was a main proponent of the national parks system and expanded their scope greatly while president. Yet for Roosevelt, and many others at the time, hunting and preservation weren't incompatible as they are often seen by urban people today. Indeed, preservation was linked to the maintenance of game, so that it could be shot (both with guns and cameras) through performances of aggressive masculinity that he then narrated for captive magazine audiences. National parks for Roosevelt served as sites for the reinvigoration of the individual and collective body, sacred spaces to be visited to confirm a white, male, heterosexual class privilege.

Perhaps because national parks act as containers for all kinds of ideas about the nation, its character, and its nature, they are also highly regulated and disciplined spaces. Animals are supervised to maintain their health and prevent disease. Fire is suppressed, set, or monitored. People are regulated through backcountry camping permits, lists of prohibited activities, and rules about interactions with the wildlife. Vision is directed to particular sights and not others. For example, Alice Wondrak Biel's (2006) book on bears in Yellowstone charts different NPS policies with respect to the regulation of nature. First, there was "aesthetic conservation" under the tenure of the first NPS superintendent, Horace Albright, between 1919 and 1929, where visitors were encouraged to feed bears and stages were built so that bears could be fed by NPS staff to the delight of onlookers. In the 1930s the number of attacks on humans had rendered bears a problem population, and the era of the "dangerous bear" emerged (Wondrak Biel 2006, 28). Yet, there was a reluctance to prohibit visitors from feeding the bears, as for many it had become one of the main attractions of Yellowstone National Park. Instead, the NPS issued warnings that bears were "wild animals," and feeding them was done at the risk of the visitor. But bear injuries continued. At the same time, the Wildlife Division of the NPS was established in 1933, which represented a more scientific than aesthetic approach to the animals of Yellowstone. This made for a reimagining of bears in the park as wild and in need of management. Indeed, in the 1950s many hundreds of bears were shipped to zoos or exterminated if they displayed aggressive behavior (28). The 1960s ushered in a return to "primitivism," in which NPS staff attempted to remake the park to look as it might have existed precontact, allowing once "tame" bears, then "dangerous bears" to become "wilderness bears" (28). Wondrak Biel asserts that from the 1970s to the present day, with its inclusion as a threatened species under the Endangered Species Act, the bear of Yellowstone has now become "imperiled" (113), as a symbol of environmental threat within a scientized view of the Greater Yellowstone ecosystem. Over the course of these many shifts in thought and policy, what remains constant is that the NPS maintained a strong hand in guiding the way the park and its nature are perceived. These kinds of regulatory endeavors—monitoring, shaping, and organizing how nature can exist and how it is perceived—have become part of the fabric of U.S. national parks.

Yellowstone National Park

Of the parks, perhaps none is as famous as Yellowstone. Its symbolism as a site of national pilgrimage is akin to the connection between Disney and childhood. Situated on a dormant super-volcano in western Wyoming, Yellowstone has achieved this degree of fame for a number of reasons. As the first national park in the United States, it was celebrated as the preeminent example of the nation's natural beauty and majesty, as well as a physical manifestation of the United States' commitment to preservation. Yellowstone is home to some of the region's most distinct natural features, like Old Faithful, Mammoth Hot Springs, the Grand Canyon of the Yellowstone, and Tower Falls, as well as the Hayden and Lamar Valleys. It also signifies a frontier landscape that once settled became much mourned in the popular consciousness and national imagination of the United States. Although the frontier myth may hold less sway now, Yellowstone has continued to remain an important element of national nature. It was made a UNESCO world heritage site in 1978. In 1988, it reentered the headlines again as wildfires decimated 793,000 acres, or 36 percent of the park (Yellowstone National Park, n.d.). This event led to a debate about whether the fires should be extinguished, pitting recreationalists and the tourist industry against scientists and park staff, but also igniting, if you will, a conversation about the primacy of science, aesthetics, private property, profit, or government policy (Delaney 1988; Schabecoff 1988). Yellowstone was again in the center of controversy in 1995 when it was the first to abide by the regulations of the Endangered Species Act and reintroduced gray wolves to the national park. Despite these controversies, or perhaps because of them, Yellowstone remains one of the most widely visited parks in the whole system, with 3,295,187 visitors in 2009 (NPS 2009b).

The story of Yellowstone is a complicated one, but not unlike other histories of imperialism, nature, and power. The Washburn Expedition of 1870 has been credited with the discursive if not the actual founding of Yellowstone. However, the prospectors, explorers, scientists, and railway owners who pushed for the founding of the first national park were not the first people there; indeed, the Shoshone, Salish, Nez Perce, Bannock, and Crow utilized the Greater Yellowstone area as a hunting range for many hundreds of years, and there is evidence that the members of the Washburn Expedition were told by Indigenous people about Yellowstone in the first place (Spence 1999). However, a story began to circulate that Aboriginal people did not, in fact, use this land. Instead, they avoided it for fear of the thermal features that dot the landscape. Although there was no corroboration for this in Native practice—which clearly demonstrated otherwise—this narrative operated as a powerful reimagining of the space that was to become Yellowstone National Park. Yellowstone was emptied of a human presence so it could be remade as a national park. Because, of course, how could this natural paradise be natural if there were people already there, living with and on the land? How could it become a sportsman's wonderland if hunters were required to compete with Aboriginal people for wolves, bison, and elk? How could it be spatially white and Indigenous at the same time? As Magoc notes, stories like Native superstitions about geysers served two aims: "first, to buttress the cultural superiority of Euro-Americans, later to support the archetypal image of unblemished wilderness we prefer in places like Yellowstone" (1999, 140). And so, this facilitated the neat (though contested) erasure of the area's first inhabitants, legitimizing Indigenous dispossession while simultaneously remaking Yellowstone as a magnet for national tourism. More precisely, the creation of the first national park guided "the transformation of the West from a space characterized by imperialist violence to a space of consumption (tourism) and production (wealth)" (Germic 2001, 96).

Consumption, then as now, was the primary vehicle to understand Yellowstone. Rather than the primeval nature of myth, settlers, most often white and male, have most often interacted with this site as a place of commerce, rather than sublime communion with nature. Created by railways, sold by hoteliers and concessionaires, refashioned again and again by NPS staff, and now a site of ecotourist adventure, Yellowstone has, throughout its long history, always been thoroughly inserted into the global tourism market.

Grand Teton National Park

Whereas volumes have been written about the history and significance of Yellowstone, very little has been said about Grand Teton. This may have something to do with its inauspicious and drawn out establishment, taking twenty years to become a unified national park. Unlike Yellowstone, Grand Teton was occupied by settlers by the time it was proposed as a national park: colonists came to the area in the 1890s, and ranchers used the grasslands below the Teton Range to graze their livestock. So, although

rivaling Yellowstone in beauty, the making of a national park here proved
more complicated, and the erasure of human presence would prove a far
more difficult undertaking. In 1916 and 1917 the NPS moved to incorporate the Teton Range into Yellowstone, a measure that was later scuttled.
The issue was put aside until 1929 when the national park was created, but
was limited to only the Teton Range and the glacial lakes, a compromise
that did not please those who felt the area to be unique and in need of conservation, like John D. Rockefeller Jr. (Skaggs 2000). Debates about the
expansion of Grand Teton National Park raged throughout the next twenty
years, characterized by concerns about loss of livelihood, control, and
"authenticity": "Ranchers worried that park extension would reduce grazing allotments; Forest Service employees feared the loss of jurisdiction on
previously managed forest areas; and local dude ranchers were against
improved roads, hotel construction and concessioner monopolies" (1).
The arguments of these interested parties effectively blocked the expansion for the next two decades.

However, the emergence of mass consumer culture after World War II
generated more of an interest in tourism. The expansion of this national
park to include the Jackson Hole National Monument was finally effected
on September 14, 1950, making the total area 310,000 acres. However,
this expansion involved accommodations that are still in effect today, like
the maintenance of existing grazing rights, the continuation of access rights,
and the agreement that there would be no more national forests, parks, or
monuments in Wyoming (Skaggs 2000). These arrangements make Grand
Teton National Park an interesting place. At one level, it seems to be a reg-
ular national park, with 2,580,081 visitors in 2009 (NPS 2009a). At another,
it is a space of labor, where ranchers still maintain grazing rights. At still
another level, it was a hunting ground for the area's Indigenous inhabitants who were removed from the land by force. But it is important to note
that in none of these iterations is it a primeval wilderness, even though it is
often consumed as such.

In reference to Yellowstone, Magoc has acerbically commented, "What
does it mean to call a place that receives three million people a year 'virgin'?" (1999, 170). A good question, and one that speaks to the large degree
of psychological work that must take place to imagine national parks as
pristine wilderness. But I prefer to ask about the consequences of imagining
Yellowstone and perhaps even Grand Teton as virgin rather than an assemblage nature–culture space. The consequences, I suggest, are romantic

nostalgia and selective remembering and forgetting. Renato Rosaldo (1989) has coined the term "imperialist nostalgia" to describe the phenomenon of longing for that which has been destroyed that simultaneously obscures the process of its destruction. In the case of nature tourism to national parks, for example, the modern anxiety about the environmental crisis has produced this kind of mourning, where there is an attempt made to recapture a lost (but of course never actually existing) wilderness. By harkening back to the supposed halcyon days of the wild United States, ecotourism reinforces the idea that nature exists in places that are separate from culture. This means that people can visit and consume national parks but never be understood as part of their nature. This concept of nature insists on the careful policing of human activity to allow only certain kinds of relationships to exist.

These kinds of relationships and practices were what the participants on my ecotour to Yellowstone and Grand Teton were hoping to experience. I now turn to the ecotour operator and the tour itself, describing the company, the participants, and the itinerary in an effort to understand how this mode of consumption works to discipline and regulate what we come to know as nature.

Natural Habitat Adventures

Natural Habitat Adventures (Natural Habitat) was founded in 1985 in Boulder, Colorado, by Ben Bressler. Beginning as a small-scale nature tour operator, Natural Habitat has expanded to offer trips to every continent, from polar bear watching in Churchill, Manitoba (their most popular trip), to "a natural and cultural odyssey" to Bhutan (Natural Habitat Adventures, n.d.1). In their trip catalogue, Natural Habitat boasts that they are the "largest privately supported international conservation organization in the world, with more than 1 million members in the United States alone" (2007, 6).[2] While the veracity of this claim is difficult to determine, and the sleight of hand in calling a for-profit nature travel company a conservation organization is difficult to swallow, there is no doubt that Natural Habitat has been noticed. Indeed, Natural Habitat recently sold the majority interest in their company to GAIAM, a self-proclaimed "lifestyle media company" specializing in an array of items linked to green living and likely to appeal to Mitchell's "metrospiritual": yoga and fitness DVDs, eco-apparel, green housewares, renewable energy products, and now ecotravel. In their acquisition of Natural Habitat, GAIAM seeks to "provide individual and group eco-travel services, allowing people to experience firsthand our planet's wildlife, wild places and natural wonders—and imparting an understanding of the critical work needed to preserve them" (GAIAM 2005, 27). And so, GAIAM did not just become the majority shareholder in a company, but adopted a vision for guiding people to save the natural world.

The Natural Habitat Adventures' Philosophy

Natural Habitat espouses a travel philosophy that necessitates that each of its "nature expeditions" must meet three conditions. First, the groups must be small to encourage interaction among travelers and with the guide. Second, the "expedition leaders" must be extremely knowledgeable about the area visited, its natural history, its wildlife, and its local culture to cater to the well-educated clientele. They must also be fit to carry people's luggage, as they did time and time again on our ecotour. Finally, the accommodations that guests find themselves in must be remote and situated within close proximity to the nature they have come to see. Natural Habitat seeks to provide a unique, personal journey connected to very particular ideas of "wild" nature.

Of course, according to Natural Habitat, their travelers are not the boilerplate "fun in the sun" tourists; rather, they seek difference because they are also unique. "We focus on providing rare opportunities for natural discovery that are out of the reach of the typical tourist" (Natural Habitat Adventures 2007, 18). While this notion of "out of the reach" may speak to the knowledge of the tour guides—or, perhaps, the wallets of the travelers—what Natural Habitat seeks to offer is *an experience* rather than just a trip. Part of this experience is linked to Natural Habitat's definition of luxury:

> To us, the word applies to the overall experience we offer our guests. To us, "luxury" means taking a helicopter flight to an ice floe to come face-to-face with a newly born seal pup, getting on a small charter plane to go deeper into the Alaskan bush, or being handed a pair of night-vision goggles so you can view wildlife after sunset. "Luxury" is venturing off-the-beaten path in the company of the most knowledgeable and experienced Expedition Leaders in the world. So, while you don't have to sacrifice comfort when you

join us on an adventure, we know that you'll appreciate our far more encompassing definition of the term. (10)

This passage speaks to an elite kind of traveler: adventurous, hardy, uncommon, sophisticated, inquisitive, discerning, discriminating, and possibly restless, one who knows that refinement is sought in the unknown, the unusual, and the exotic rather than in sumptuousness. In this way, the eco-traveler is promised a different kind of opulence, a wider abundance of sights and senses to experience. It is because of these qualities that the Natural Habitat tourist knows the value of saving the special places on earth and this supports the company/organization's "commitment to conservation."

Commitment to Conservation: Natural Habitat Adventures and the World Wildlife Fund

In many ways, Natural Habitat Adventures is an ecotour operator much like any other, featuring visually stunning catalogues awash with romantic views of nature, attractive Web sites replete with appealing animals, trips that seem to conflate nature and culture: in short, a corporate enterprise with a sophisticated marketing team that works to bring customers in. But it is also more than this. Natural Habitat has been named as the World Wildlife Fund's (WWF) Official Conservation Travel Provider.[3] What this means practically is a few things. First, when WWF members choose a Natural Habitat trip, a portion of the proceeds is donated back to the WWF; second, that Natural Habitat will sponsor a one-year membership for those who are not WWF members but would like to be; and, third, that the tour guides are aware of WWF efforts in conservation, and will cross-market their work. According to one of the guides I interviewed, Natural Habitat is one of the WWF's main financial supporters, presumably contributing to the $7 million in funds the organization received from corporations last year, though no specific number is provided (WWF 2007). But this is also a reciprocal relationship. WWF shares their mailing lists with Natural Habitat and advertises their trips on their Web site, providing a ready-made clientele of nature lovers for this ecotourism operation.

The naming of Natural Habitat as the sole Conservation Travel Provider for WWF, an organization of some 1.2 million members in the United States and over 5 million worldwide, means more than simple practicalities. It also signifies that the service Natural Habitat provides is somewhere outside the stain of mass tourism and capitalism, much like the Environmentality Program attempts to do for Disney. And so, Natural Habitat can claim to be more than a moneymaking company. Instead it might be read as a hybrid, neither completely capitalist nor fully nonprofit but somewhere in between, a conservation organization that can make money by doing so, a sustainable development strategy at its finest. Thus, Natural Habitat can purport that by traveling the tourist has participated in wildlife conservation. As one former Natural Habitat traveler asserts in the company's catalogue, simply visiting a place is not only personally transformative, it actually works to stave off possible environmental threats: "The gifts of beauty and grace I received from my trip to Mexico to witness the butterflies will be a part of me forever and I will be changed in some way by being present to this miracle. The entire experience and my participation from start to finish, makes me feel that I have—in some small way—contributed to the survival of these magical creatures and the local areas we visited. WWF has chosen their partner wisely" (N. Shelby, Four Time NHA Traveler, qtd. in Natural Habitat Adventures 2007, 7). So not only is Natural Habitat relieved of the problems that a more complicated analysis of green capitalism might compel, but the ecotourist is safe in the knowledge that their expenditure has served and saved nature.

In recent years, however, Natural Habitat has been forced to recognize the environmental damage that tourism, even ecotourism, inflicts on the natures they purport to save, even quoting estimates that travel-related activities account for almost a third of "greenhouse gas emissions that affect global warming" (Natural Habitat Adventures, n.d. 4). In concert with WWF, Natural Habitat has instituted a carbon pollution reduction program where all travelers can simply enter their destination information into an online carbon calculator and they will be given a particular figure of "how much money it will cost to prevent the same amount of gases from being emitted elsewhere through investments in alternative energy projects in the developing world" (Natural Habitat Adventures 2007, 9). The client can then donate the appropriate sum to Natural Habitat, which it then assigns to a particular project. In any case, the Natural Habitat customer is not dissuaded from traveling but rather is encouraged to participate in an easy program to offset the greenhouse gases associated with the trip, "allowing you to participate in nature expeditions that are 100 percent positive" (9)! Ecotravelers are offered the opportunity to use their wealth to buy a way out of responsibility for environmental problems.

Hidden Yellowstone and Grand Teton

Natural Habitat's exposition of its travel philosophy in the United States is so romantic, so nationalistic, and so emblematic of the kinds of language used to describe nature in ecotour operations that it is worth quoting at length:

> It can be easy for a world traveler to overlook the *natural splendors* that exist here at home. Because we believe that America's *unspoiled expanses of wilderness* and its magnificent wildlife rival those found anywhere else on Earth, we *pursue* domestic travel experiences that are *remote and secluded,* and close to the wildlife, just as we do on an African safari. Our professional *Expedition Leaders* escort *small groups* of travelers to the most remote parts of America's famed wildlife refuges. We leave the crowds behind at the typical tourist hotspots and *venture deep* into the heart of these protected lands, finding ourselves within feet of grizzlies catching spawning salmon in a clear Alaska river, watching a pack of wolves hunt for food on an early winter's morning in Yellowstone, or quietly witnessing the sunset over the colorful ridgeline of the Grand Canyon. Join us in America for a *true safari experience.* (Natural Habitat Adventures 2007, 46, emphasis in original)

Of course, one might suggest that this is simply marketing, meant to sell exotic vacations to urbanites hungry for a bit of respite, or perhaps to Americans who, in the context of the so-called "War on Terror," are now afraid to leave their own countries. But the fact that it is marketing is exactly the point. Its reliance on the rhetoric of romantic and exotic nature, its invocation of explorer/traveler who is willing to risk getting off the beaten track, its use of evocative language to construct a scene one might watch in a nature film, and its appeal to a nationalistic sense of nature make this passage one that is freighted with signifiers that white Americans can understand and relate to with little conscious thought. This passage can draw would-be ecotourists in because it makes manifest culturally significant metaphors and signs of power: nation, purity, and wild(er)ness all find their home in this brief excerpt. Indeed, one is reminded of the Disney claim that the Kilimanjaro Safari ride is Africa without the mosquitoes; for Natural Habitat's part, it is the safari experience (close contact with desirable animals, small groups, unspoiled wilderness) that they bring to the United States. Domestic nature is recast as exotic, and Americans can perform the nationalist impulse of seeing "their" nature.

One part of this story of pristine nature in the heart of the postindustrial West is the idea that something has been lost or obscured and can be found. Due to the expertise of the "expedition leaders" and the resilience of the guests, Natural Habitat can find secluded locations even in Yellowstone National Park (Natural Habitat Adventures 2007). In what are marked as "journal entries" in the Natural Habitat catalogue, a former guest of this trip notes:

> When I was a child, I got lost in a park. I didn't like it much. But now, 40 years later, I search out these types of experiences. In Yellowstone, getting lost isn't as easy as you might think. It's a popular place, and I think lots of people have the same idea. That's where my guide helped. He led me and a co-traveler onto a vacant trail north of Old Faithful. We ambled and talked about the geysers. We ambled some more and talked about the bears, moose, and eagles. After a while, I lost my bearings. Our guide knew the way, but what's important is I could get lost in a park again. (54)

The desire to get lost, to feel secluded and possibly at risk (all within the safe and protected confines of an ecotour), seems puzzling, at least at first glance, particularly in a national park that receives some 3 million visitors each year. Indeed, this feeling is completely disconnected from the actual experience of the ecotour and the park more generally. Our guide (2006) asserted, "90% of people who visit Yellowstone don't go a mile from their camp. Most people who go to Yellowstone get the road experience, rarely straying from the boardwalks and roads." He indicated that this was compounded by the age of the typical Natural Habitat ecotourist, who unlike the eco-adventurers described by Braun (2002), Farley (2005), or Waitt and Cook (2007), seek less strenuous kinetic experiences. Indeed, as Gilbert (2004) has pointed out in other contexts, this kind of trip straddles the fence between rustic and luxurious, where the travelers take short jaunts into the wild with their guides and then return to the safety and comfort of their hotel rooms. So, how can we make sense of this desire to be lost that so clearly seems to be virtually impossible on trips such as these? I think it must be read as a performance of mourning as similarly described by Braun (2002)—an attempt to shed modernity by grasping at the pre-modern, while at the same time dragging the trappings of the modern along with them to the experience. Remoteness, seclusion, the possibility of being swallowed by nature, but, of course, pressing ever onward: these are the things that both threatened and titillated the early explorers. In the present, this can be read through the lens of a neo-imperial desire to consume the unknown: to experience that which is rare and undiscovered. These, then, are the psychological tools that allow the ecotour to "Hidden Yellowstone and Grand Teton" to function.

The Itinerary

Our itinerary for this trip is detailed on the map below. We began and ended this seven-day tour in Jackson, Wyoming. From there, we spent some time in Grand Teton National Park, then journeyed up to Yellowstone National Park to see Old Faithful, Mammoth Hot Springs, and the Lamar Valley, staying overnight both in the park and in Cooke City, Montana. Each day of this trip is explored below, drawing together the theoretical insights of governmentality as well as approaches to tourism to understand how this ecotour governs the senses—aesthetic, intellectual, physical, and affective.

Day 1: Jackson—The Gear Fetish and the Wealthy Environmentalist

Jackson, Wyoming is, without doubt, a rich town. Set amid the glacial lakes and peaks of the Tetons, this picturesque village is something of a resort town, offering wildlife viewing, rafting, mountaineering, hiking, and dude ranches in the summer as well as skiing and snowboarding in the winter. Former vice-president Dick Cheney, as well as celebrities like Harrison Ford and Sandra Bullock make their second homes in Jackson (Banay 2006), and the main street is littered with quaint boutiques and ski shops. Nature and profit coincide in the spectacular setting of Jackson, which acts as a bubble insulating residents from the more complicated picture of land politics in the West. So, the drive through Wyoming reveals a primarily ranching landscape, dotted liberally with trailer parks and, in locales like those outside of Cheyenne, the new exurban "ranch-lets" with subdivision homes set on one or two acres of land. Moreover, it seems that those who service the tourist economy in Jackson can scarcely afford to live there; our two guides on the ecotour lived in Kelly, a small town some twenty kilometers away, indicating that housing in the resort town has become too expensive. The wealth of Jackson makes it a suitable starting point for this ecotour, where nature meets comfort, privilege, and luxury to produce a commodified wilderness experience.

The ecotour participants arrived at the well-appointed yet appropriately rustic-looking hotel, replete with elk antlers, large wood furniture, and paintings of local wildlife to make one feel that we had now entered the mythical West of which some of us had dreamed. This was the hotel where we would begin and end our journey. We convened in the lobby to go for dinner and to have the first briefing by our guide. This briefing supplemented the information that we had received from Natural Habitat before departing, which included books to read, clothes to purchase, and what to expect

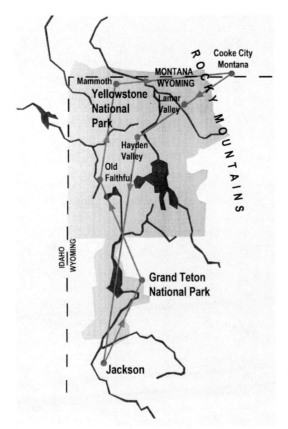

Hidden Yellowstone and Grand Teton. A map of the itinerary taken by the ecotour through the national parks. The ecotour began and concluded in Jackson Hole, Wyoming.

as part of the trip. The briefing provided more specific information and involved a discussion of the itinerary. We received maps of the parks as well as a checklist to mark off the animals we saw on the trip.

Eager with anticipation, we chatted about the reasons for embarking on this trip. What appeared to drive U.S. travelers to embark on such a trip was a nationalist impulse to "see our own country." Of course, the motive was hardly new to travel in the United States; Schaffer (2001, 103) has asserted that "scenic patriotism" was a main impetus behind the birth of the U.S. national park system. The feeling endures, then, combined with the desire participants expressed to see the majestic nature of the West, encounter its animals, and come away with new experiences.

From the hotel, we moved on to dinner and the opportunity to get to know one another better. As we introduced ourselves, it became clear that the participants on this Natural Habitat trip bore more than a passing resemblance to the typical ecotourist described earlier in this chapter, but also were different in important respects. There were eight participants on the ecotour, with one guide and one logistical organizer whose role was to arrange accommodation and meals. Seven were women, two of whom were traveling alone. There was one married couple and two friends who had met on a previous Natural Habitat trip. All were white. They were largely an urban crowd, living in or in close proximity to New York City, Toronto, Miami, and San Diego, with one traveler from Iowa. Each of the participants had some form of postsecondary education and held a range of positions: office administrator, urban planner, volunteer fundraiser, and landscaper. We had three retirees on the trip as well—a human resources manager, a teacher, and a secretary. Almost all of the participants had gone on an ecotour with Natural Habitat Adventures in the past, and for one participant this was her tenth trip with the company. Given the price range of these trips—with sea turtle watching in Mexico at the lowest end for U.S.$2,195 to $21,833 for a private cabin to recreate Ernest Shackleton's voyage to Antarctica (Natural Habitat Adventures, n.d.1)—there is obviously a particular class dynamic at work here. The travelers were also generally speaking much older than the average ecotourist cited above (between thirty-five and fifty-four), with one participant in her seventies and most in their sixties. According to the guide, the participants on this trip were typical to Natural Habitat: "Older folks with disposable income and time. They are mostly between 50 and 65. More women than men go on these trips and often widows." When I asked him the degree of ethnic diversity, he indicated, "Oh it's all white people. We have to fill out these forms . . . on participants. I get so used to filling out Caucasian that I often miss the occasional Asian who signs up." And so, we have a profile of who goes on Natural Habitat trips: white, female, urban, older than the typical ecotourist, looking for the comfort and security of a guided trip, equipped with disposable income, and interested in wildlife watching.

The gendered composition of our tour group stands in opposition both to the average tourist described above as well as the history and contemporary readings of adventure-based ecotravel that have focused on the masculine performance inherent in the practice. Indeed, harkening back once again to Roosevelt, travel into the wild was predicated on the recapture of a white masculinity lost through both the feminization of the city and its racial decline. While modern adventure travel may have lost some of this explicit reference to race, gender, and sexuality, some tours, particularly those that are physically taxing, associated with risk, and competitive, continue to carry this freight (see, for example, Braun 2003; Erickson 2003). However, while the Natural Habitat excursion may be contaminated by the trace of its origins (for example, the emphasis on getting lost in the wild or exploring the unknown), there is also something a little more complicated at work. The gendered composition of the ecotour speaks to a practice that subverts its beginnings, in which, at least to some degree, the emphasis on masculinity has leaked out. This ecotour is about *looking at* animals rather than climbing mountains, for example. Unlike some of its other manifestations, this trip is risk averse: it doesn't involve tests of physical ability, potentially dangerous contact with animals, or competition. And unlike big game hunting, which shares a similar focus on animals, the primary technology of power is the camera rather than the gun (see more on this below). The fact that the majority of people on this kind of trip are women speaks to this alternate construction of ecotourism. Thus, there is a microphysics of power at work here, but one that cannot be clearly sorted into the categories of either masculinity or femininity. More precisely, the practices on this ecotour and the people who participate in them resist easy classification. As we will see, from a fixation on gear to the watching of wolves, race, gender, and class infuse this ecotour in complicated and sometimes contradictory ways.

Given that so many of my fellow travelers had been on these trips before, what became immediately apparent was that there were experts among us: ecotourists who knew their way around a safari or Arctic expedition.

This expertise was evident through the acquisition of, and discussion about, the gear required to be a real ecotourist. Now, this is not the gear of extreme or adventure-based travel—there are no avalanche beacons or climbing axes here. But gear is also no longer solely the purview of mountain climbers in Nepal or trekkers in the Atacama. As John Tierney indicates in his *New York Times Magazine* article,[4] the celebration of gear had grown to include a once-foreign client base:

> In America, the epicenter of the boom, the number of adventure-travel outfitters has tripled in the past decade, to more than 8,000. Their clientele is increasingly older (the average age is about 50) and more heavily female. Sales of outdoor equipment have quintupled since the mid-1980s, to $5 billion a year, partly because more Americans just want to dress like explorers (sneaker companies are hurting because fashionable urban youths have switched to hiking boots), but also because so many are going into the wilderness. (1998, 21)

Tierney, an ambivalent and often grudging actor in adventure tourism, simultaneously laments and revels in his desire to buy the newest, trendiest, and most expensive items to reenact the imperial contest. Indeed, no one seems immune to the "gearhead" disease, and the participants on my ecotour were no exception. This first night and throughout the trip there were long and expert discussions of the merits of merino wool versus silk long underwear, the importance of Gore-Tex outerwear, and how to select the right hiking boots (keep in mind that we spent the majority of our time riding around in a van equipped with hatches in the roof you popped open to see and photograph animals). While style, utility, and "authenticity" mattered with respect to gear, price was rarely mentioned. Beyond the clothes, my fellow travelers carried a small fortune in cameras, often rivaling the cost of the trip. Gear was clearly an important, if not central facet, to the identification as ecotraveler.

Patagonia, The North Face, Columbia: each of these brand names served as markers not only of wealth but also of the ethical impulse to do right with said wealth. Jackets, fleece, backpacks, hiking boots, long johns, and cameras were fetishized—made into more than themselves and imbued with all manner of meanings other than their use. The fetishistic character of gear in nature travel works to alleviate anxiety and dread about

what is seen as the loss of the wellspring of modern life: nature. This consternation is mediated through the act of purchasing a trip to a supposed
remote wilderness and is crystallized through the complementary yet pre-
determined act of outfitting oneself for this trip. All manner of commodities—tours, clothes, cameras—are invested with values that are separate
from the labor and materials that brought them into being. Gear works as
a totem, its purchase cleansing the product of its capitalist manufacture
(although it is essentially shopping for a brand) while simultaneously re-
assuring the purchaser of the ability to go into the wild.

The gear fetish performs another, yet related, role: it ascribes a particular identity. It is a form of self-fashioning. The act of outfitting—of considering all the options, of selecting the right items, and of purchasing
them—offers some of the materialities of subject formation. In my esti-
mation, the subjectivity claimed has two dimensions. First, like those produced in the museum and at Disney, the subject–position asserted is that
of environmental citizen: in the act of outfitting, one quite literally dons
the markers of environmental awareness, experience, and action. Hiking
boots and Gore-Tex jackets signify a person who likes to be out in nature,
is knowledgeable about it, and who is interested in saving it. Like eating
organic or driving a hybrid car, ecotravel and its attendant rituals stand as
evidence of a particular sensibility: a commitment to environmental responsibility and green citizenship.

But while these technologies of the self can be read as markers of environmentalism, they are not necessarily indications of the choice to live a
life less complicated by capitalism. In fact, quite the opposite is true as the
green impulse becomes more deeply intertwined with commodification.
Gear is a growth business as more and more people seek to sport the accou-
trements of outdoor activity. The gear fetish, and the ecotourism that
accompanies it, not only signify environmentalism, but also wealth, or at
least disposable income. Perhaps the message here is that one needs money
to care for the environment. And these items have a built-in obsolescence,
making the quest to keep up to date with the latest designs an even more
expensive proposition. As Tierney has indicated, none of these items come
cheap, and because of this, there is pleasure and play in the ritual:

> Even I, who dreaded the prospect of an Arctic trek, loved
> shopping for it. . . . I happily spent hours in . . . gearhead bazaars
> fondling smooth layers of Capilene and Polartec, agonizing

between Gore-Tex and Supplex, picking out gloves and gauntlets, glacier sunglasses and a chronometer with a built in thermometer, barometer and altimeter that I absolutely had to have. As I walked across Central Park in April to train on a cross-country ski machine at my gym, I exulted in my monstrous new leather-and-rubber LaCrosse pack boots guaranteed to 100 degrees below zero. How manly! Take that, Nature! Just to be safe, I also bought a pair of Steger Mukluks [boots] for $135. (1998, 22)

In the wearing or use of gear, there is not only the performance of an environmental subjectivity but a moneyed one as well; these people have not only the inclination but also the means to be better environmental citizens. And they choose to perform their environmentalism outwardly through the expenditure of money. As our guide indicated, sometimes it is not even so much the wildlife that ecotourists are interested in, but the display of wealth: "These trips become cocktail conversations. Some participants have no interest in looking at the wildlife but rather spend their time in the van talking about their previous trips." Certainly, this is not the case for all or even most of the participants, nor does it preclude other means of environmentalist action—perhaps they are involved in animal rights organizations or green activism at home, though none of the ecotourists mentioned this kind of activity—but it does demonstrate one of the ways that nature, profit, status, and power come together in the practice of ecotourism.

At the same time, as Tierney's pithy quote above indicates, more than just environmentalism or the expression of class, this is also the garb of adventure: the ecotourists don the apparel of an explorer to render palpable the trial to come. The focus on exploration seems to be a complex one, given that I have argued that the trip itself was risk averse. Why, then, was there a need for gear that could take one into the Arctic Circle instead of a rather cozy van? Here again, I think, we find the trace of the masculine impulse on this ecotour. Of course, shopping and its attendant rituals have long been cast as a feminine endeavor. However, the articles purchased speak to something else: a buy-in to the fantasy of vigorous engagement with the wild. That it is women that engage in the practice makes it both more complicated and interesting. It seems to me that the gear fetish performed as part of this journey is certainly more subtle; as mentioned above, its focus was on warm clothes or the ability to take stunning pictures. But the display of this gear was also competitive—assessing the quality of long underwear or the capacity of cameras had a definite undercurrent around who had the best gear, which is linked to how much it cost. And so, the gear fetish on this tour offers a window into an environmentalism made manifest through consumption as well as the complex character of self-fashioning that works through both class and gender.

Day 2: Grand Teton—Animals Eating / Eating Animals

The next day the trip began in earnest as we entered Grand Teton National Park to begin viewing wildlife. We spent the day mostly driving, stopping to get out to look at wildlife when the guide deemed it appropriate, and taking pictures of elk, bison, and rutting moose. This was typical of the rhythm of the tour, where the guide would tell us what time to get up; when to exit the van; direct us to interesting sites; arrange breakfast, lunch, and dinner for us; and, as we drove, provide detailed information on the animals of the Greater Yellowstone ecosystem.

There was a feeling of anticipation in the van, excitement about the possibilities of seeing wildlife that for most of us had not been seen outside of a zoo. As we came upon a large herd of bison, one of our bus-mates exclaimed, "Where are the Indians?" Calling up imaginings of the frontier, this ecotourist seemed to make this throwaway remark in both jest and lament. The guide explained that that there were not many Native Americans in the area, and historically the Grand Teton area was used only for summer or winter hunting. In one swift move, both the participant and guide seemed to relegate Indigenous people both to the past and as part of nature. Of course, this taken-for-granted story is a contested one. The history of this area, as explored above, indicates that Indigenous people made good use of what is now a national park. While the U.S. Census Bureau indicates that in the year 2000 only sixty-seven Native Americans lived in Jackson (U.S. Census Bureau 2006), the Wind River Reservation is actually in close proximity, with some twenty thousand Eastern Shoshone and Northern Arapahoe living on 2.2 million acres of land; clearly present-day peoples who did not die out with the last bison herds chasing across the Great Plains. Moreover, the discursive association between Indigenous people and buffalo means that the two can be conflated and simultaneously made into a thing of the past. This, of course, is a prevailing theme in much of the ecotourism, particularly in, but not limited to, the Global South, where Indigenous peoples are as much on display as is wildlife or scenery. The move to make present-day Indigenous peoples both anachronistic and part of nature makes a certain kind of awful sense in this place. Here, in these sites of national nature, there can be no recognition that this land isn't "American," that there are other claims and histories. Instead, ecotourists accept their presence as part of an evolution of the nation, where progress dictated the extermination of both animals and Native Americans. And so my fellow ecotourist's remark, as well as the guide's response, firmly position Indigenous people in a regrettable, but necessary, past.

The quest for difference is an important element of any ecotour. As Jane Desmond has reminded us in her discussion of the connections between cultural and natural tourism, "both continue to constitute a contemporaneous sense of what their viewers are by showing them what they are (supposedly) not" (1999, 144). *Unlikeness,* then, is part of what ecotourists seek when they gaze upon animals: the self is constituted through contemplation of the other. But this quest does not end with the visual and metaphorical capture of the "last great species of the American West." The encounter with radical otherness also presumes that the modern subject can move back and forth in time. So, as Desmond argues, ecotourists engage in an imaginative as well as physical journey, "Crossing from our world into theirs provides a fantasy of returning to our origins, or becoming part of the natural world ourselves, at least for the duration of the visit" (190). Once the ecotour is over, the modern life of the ecotourist is reinstated. So the encounter with unlikeness assists the ecotraveler in negotiating their own sense of self, positioning them in the world as both part of and opposite to nature, depending on the context.

We spent the remainder of the day chasing after this radical otherness in our vehicle (in fact, we almost hit a deer), as well as taking a small hike to see a moose with her calves. In the evening, we met for a meal. What was striking here were the easy and unspoken distinctions that people had drawn in their minds between rare and familiar, wild and tame, real and unreal. After having gazed upon bison on the range, two of our number ordered bison burgers at dinner. Confusion must have shown on my face. Were these not, in fact, the same animals? One of the burger-eaters said sheepishly, "Well, these bison are raised for this purpose on a farm." And so, the reason for this differentiation came into view. One carried the majesty and cachet of being "wild," the other burdened with the banality and incarceration of industrial agriculture. Of course, an examination of the history of animals at national parks like Yellowstone reveals that the hard lines drawn around wild versus domesticated need to be blurred. The intense management of wildlife for display, the slaughter of predators so that animals like the bison could be seen in large numbers (and their over-flow hunted), the reintroduction of wolves to reestablish a whole ecosystem: all of these actions belie the moniker of "natural" when discussing animals at national parks and speak to the biopolitical enterprise of park management. Even the flora has been stage-managed differently at different points in the history of the park.[5] While the regulation experienced by animals at a national park is quite different from, for example, the disciplinary and often torturous circumstances endured by animals at factory farms, there is also little freedom here. The rationality of making distinctions between animals of the same species because some are "wild" while others are "tame" seems silly in the face of such intense management—free and captive are never as separate as they seem.

However, these kinds of distinctions do have political effects. The division between wild and tame allows us to venerate one while erasing the importance of the other. In only counting the animals we encounter through

"Wild" Bison, Yellowstone National Park.

a wilderness experience *as animals* rather than *as food,* we separate this experience from our everyday lives. It becomes unique, special, and uncommon. Thus, the ecotourist can mourn the loss of habitat for wildlife while simultaneously not making the connection that intensive livestock production consumes both the land and resources of their wild cousins. Livestock is made part of culture, while wildlife remains resolutely part of nature. This disconnect prevents a more radical critique from emerging; instead, this understanding produces a particular kind of political engagement: supporting organizations like the Nature Conservancy or the World Wildlife Fund, engaging in outdoor recreation, and, of course, going on ecotours.

Day 3: Yellowstone—The Truth of Science

The third day of our trip began by our leaving Grand Teton to make the journey to Yellowstone. We stopped at lunchtime for a picnic in the snow. Over lunch, our guide began to explain the fire ecology of Yellowstone, for as we entered the park many of my fellow travelers expressed an interest in the lodgepole pines that still bore evidence of the 1988 blaze. He explained that the NPS practiced fire suppression for most of the park's history. By the 1970s, however, the NPS had come to understand fire in a different way, adopting a natural fire policy in 1972. Drawing on historical and sedimentary records, the NPS heeded the arguments of ecologists about fire as a natural process and began to recognize its ecosystem benefits. Now fires are allowed to burn in a controlled manner and sometimes the NPS sets fires themselves.

However, the history of fire ecology in the national parks, although interesting, is not my main concern. Rather, what I am keen to explore is the way this example and others work to establish science as an important element in the making of environmental subjects. The discussion on the merits of fire ignition is just one example of how on almost every occasion possible our guide, who was also identified as a biologist, described the zoological characteristics and natural behaviors of the animals we watched. He spent long periods talking about ecosystem health and the necessity of understanding the interconnectedness of life. A discussion of Old Faithful led to a scientific explanation of geysers and fumaroles. In fact, the entire trip was suffused with an attention to the scientific basis of nature, much like at the American Museum of Natural History.

When I queried the guide about this tendency that seemed to undergird the ecotour, he suggested that understanding the science of ecosystems is key to building better environmentalists:

> You have to get across the importance of science and ecology, and that interconnectedness is what makes an ecosystem sound. If people don't know this then they don't know what to save. . . . But if you can't understand how ecology works, you can't understand the larger issues. For example, Yellowstone is a small park, although most people think it is big. Without understanding the science, you don't have a sense of how unsustainable Yellowstone is, especially in the winter when animals almost starve and then leave the park and enter our screwed up political environments. Usually what happens is these animals that are protected in Yellowstone get killed or captured when they go outside the park. Bison are killed when they leave the park for no reason other than wanting to eat grass. What we want is people making connections between what happens now and what happens three months from now.

For our guide, then, becoming a true environmentalist necessitates not only love for nature but also an understanding of its dynamics. And so, in large part, this ecotour was meant to operate as a kind of project in environmental education, one that is in service of profit. As our logistical coordinator noted, "We want people to understand what's going on rather than just seeing the animals. Really, the entire tour is based on teachable moments in terms of responding to what people are curious about." Emphasizing the import of science—attempting to teach people to become better environmental citizens—was what the guides felt made the tour significant.

It seemed that what this ecotour was attempting to do was walk a line between the affective and intellectual dimensions of the desire to encounter pristine nature. On one level, people are driven by the emotional response, or perhaps affective desire, to be in the presence of radical otherness—a nature that they can neither understand fully nor conceptualize. However, it seems that this remains insufficient. What these travelers also need is an education to be dispassionate, objective observers of natural processes. Perhaps this, when coupled with the right gear, makes one the kind of environmental subject that can save the last of a remaining nature.

Day 4: Yellowstone to Cooke City, Montana—Photography and the Visual Grammar of Nature

Day 4 of our trip began with a walk around the Old Faithful area to see the thermal features that make Yellowstone so famous. Along the way, we ran into a bison and two grizzly bears picking their way through the geysers to reach the forest beyond. Our guide chatted about the ecology of the paint-pots, fumaroles, and springs, as well as the microbacteria that can flourish in such conditions as we wandered back to the hotel to pack up and move on again. From there, we journeyed to Mammoth Hot Springs, the aptly named terraced limestone spring. Across from the springs are the remnants of the old Fort Yellowstone (now the park headquarters) and a museum/visitor center/gift shop. It is here that visitors come upon the elk of Mammoth. Always in residence, these elk seem to present themselves to the gaze of the viewer, or perhaps have become so used to humans in their habitat that they seem supremely unconcerned. Except, of course, when they are rutting, which they were. Hence, we were strongly warned not to exit the vehicle. After watching the elk fight for quite some time we got back on the road, headed for Cooke City, Montana, our destination for the evening. On the way, we stopped at the Lamar Valley to see if we could spot some of the wolves of Yellowstone. Through high-powered spotting scopes, we were able to catch a glimpse of the elusive animals, which we encountered again on day 6. We moved on to Cooke City, to dinner, then to bed.

What these moments had in common was that each landscape, each attraction, and each animal was thoroughly photographed. Like a horde of nature paparazzi, we moved from sight to sight, snapping photos of charismatic megafauna to commemorate the occasion and to share with friends and family back home. Indeed, I began to wonder how different these animals might look if viewed with the naked eye rather the lens. Of course, this is a phenomenon not limited to this trip, or even to ecotourism, but instead has been a defining feature in the way mass tourism has developed in the twentieth century. Urry (2002) has suggested that much tourism has become organized around the collection of photographs of places deemed beautiful, unique, or significant, and this in turn has accelerated an already present commodification of memory. However, I contend that this does more than turn landscapes into commodities. It provides us with a compelling technology of vision with which to apprehend nature.[6] If we turn to theorists like Sontag (1978), we can see that photographs not only commodify, but also govern through their transformation into product.

Sontag (1978) argues that the modern practice of photography works to structure our ways of seeing, teaching not only the individual photographer but the population at large the centrality of vision and the ability to judge what has meaning as an artifact worth collecting and, consequently, what does not. Moreover, she contends that photography shrinks the world, making it fit within a repertoire of images that can be knowable, catalogued, and assessed: "the most grandiose result of the photographic enterprise is to give us the sense that we can hold the whole world in our heads—as an anthology of images" (Sontag 1978, 3). By dividing the world into that which is worth seeing and hence known and reproduced, and conversely that which is not worth seeing and hence unknown and obscure in its prosaism, the photographic representation of tourist landscapes offers a regime of regulation where the images we see circulated again and again become the truth of a place, its nature, and its people. In this way, as Sontag suggests, "[t]he photographer both loots and preserves, denounces and consecrates" (64). Ansel Adams's government-funded project of photographing spectacular nature in Yellowstone and Grand Teton provides an excellent example of this effect. His images of Grand Teton, the Snake River, Yellowstone Falls, Old Faithful, and the Fountain Geyser Pool have become the iconography that amateur tourists seek to replicate on their visits to these sites. Places become destinations, sights, scenic drives, highway turnouts: in short, an amalgamation of images that can be taken home and unproblematically shared. In short, these places become memories, and photos act as mementos of the ecotour experience. This allows people to travel imaginatively, to know the essence of a place without ever having been there. In this way, photography invents and reinvents our world, stabilizing parts of it, or at least their representations, for easy consumption.

The need to record—to fix in space and time—has a particular resonance in ecotours as they are heavily inflected with mourning, nostalgia, and romanticism. Generated in part through anxiety over the perceived loss of nature, ecotours have become spaces where, at least temporarily, travelers can perform a return to a previous time, where nature was not under threat. Of course this return is impossible, but the viewing and capturing of landscapes and animals through the lens allows for this anxiety to be relieved and control over an ever-changing nature to be reasserted; the photograph fixes in time a fleeting nature under threat. As Urry suggests, photographic practice involves the inscription of mastery: "The photographer, and then the viewer, is seen to be above, and dominating, a static and subordinate landscape lying out inert and inviting inspection. Such photographic practices demonstrate how the environment is to be viewed, dominated by humans and subject to their possessive mastery" (2002, 129). What Urry does not go so far to say, however, is that in this act both subject and object are regulated—governed—and nature becomes one more sight to be captured, circulated, consumed, and reified when one shows the pictures to family and friends.

Moreover, nature photography has been linked to the safari or expedition. Because the camera functions through a "penetrating" gaze, it has been likened to a gun. In the case of the American Museum of Natural History, as explored by Haraway (1989), Carl Akeley used both camera and gun to capture and represent the animals of Africa in the pursuit of science: to document a disappearing nature and to represent it for a metropolitan audience. However, as both Brower (2005) and Dunaway (2000) assert, the historically omnipresent rifle gradually vanished in such endeavors, replaced by the camera as the tool for collecting specimens. Teddy Roosevelt, for example, argued that conservation dictated such a supplanting; photography offered some of the same pleasures as shooting in terms of a vital encounter with the wild, and could be equally as important in building a white, masculine "virile citizenry" (Brower 2005). Sontag contends that this replacement is the sign of a psychic sea change: "Guns have metamorphosed into cameras in this earnest comedy, the ecology safari, because nature has ceased to be what it always has been—what people needed protection from. Now nature—tamed, endangered, mortal—needs to be protected from people. When we are afraid, we shoot. But when we are nostalgic, we take pictures" (1978, 15). When we photograph instead of shoot, then, nature has been brought under the (perceived) dominion of humans, and its passing is to be lamented and documented in equal measure. Paradoxically, the desire to chronicle this fast disappearing nature works to hasten its demise: the more people who trample through the bush to get the perfect picture of bison on the range, for example, the more their habitat is degraded. The urge to encounter "wild nature" perversely makes it less and less likely that such nature can ever be found.

Turning back to the travels of the ecotour group, these dynamics can be seen at work, although the phallic and masculinist references are complicated once again by gender. Snapping rolls of film with high-powered lenses or freezing scenes with digital cameras, the main preoccupation of this trip was taking pictures. The ecotour was structured to provide the best opportunities to have this visual experience of nature: a short side-trip to the most photographed barn in the United States, stopping at almost every scenic outlook in the Tetons and Yellowstone, lingering at Old Faithful, using the telemetric information of NPS researchers to help us find animals. What is included is what is deemed majestic and noble, aesthetically pleasing, unique, exceptional but familiar—icons that sum up "the West" and the "frontier," while signifying the United States as a nation with enough foresight to cherish and protect its natural treasures. What is excluded are the more mundane elements of any trip to a national park: the more common kinds of animals like raccoons, squirrels, and shrews; the traffic jams that occur as a result of an animal sighting, a symbol of the technology it took to get there; or the other people who visit or work in the park, whose presence might challenge the notion that it is primarily a natural place. These aspects of national parks do not merit a record. And so, ecotourists see what their imaginations and anticipations of the trip

Moulton Barn near Grand Teton National Park, named the most photographed barn in the United States.

told them they would see, because it has been visually represented by so many images; the photographic representations of these places make the trip a quest to reproduce what has already been consumed. In this sense, the eye is already governed through an aesthetics of power.

I think a comment by our guide is worth quoting at length because it illustrates the "point-and-click" nature of this vacation:

> But some participants, all they care about is getting the picture. But what they don't realize is that none of the close-up pictures of wolves, for example, are wild. They are all taken in captivity. *If they don't see it in their eyepiece it doesn't exist.* A big percentage of the people we get are photographers. But if you look at the polar bears in Churchill, they get too close to the van. This is for two reasons. First, they are not fearful creatures; they are inquisitive and pretty bored so they will come up to the van. But also, they are habituated to humans. Either way, people like these trips because they can get close ups of the animals. But it is not especially natural. If you went up to the high Arctic, the polar bears often run from humans. The animals in Yellowstone are habituated to humans as well. The ravens, the elk in Mammoth, they are getting something from humans that they don't normally get. That's why you can get so close. (emphasis added)

There were few concerns about the ecological impacts of this quest for the perfect picture, and little questioning about the fact that some of the animals photographed, like the raven and elk, seemed as used to humans as the house pets they had left at home to make this journey. Remarkably, there was no consternation about wildness, or its seeming lack. Rather, the ability to get close, to get a good shot, to see an animal through a lens, provided the real thrill. Like hunting expeditions of the past, the number and quality of animals captured seemed in some cases to count more than the experience of seeing the animals themselves.

This is further illustrated by a moment that spoke volumes about the visual nature of this trip, and the collecting of animals like curios. Although it did not happen on day 4, I think it merits mention here, given the tenor of my discussion. On our ride back to Jackson on the last day of our trip, we stopped at a general store (as was our usual practice, in order to use the facilities and purchase all manner of gift items and keepsakes).[7] While a few of us waited around the van for the others to return, one of my co-travelers showed us some postcards she bought that depicted wolves. We had just watched wolves that morning, but it was almost impossible to photograph them because they were too far away. By way of parlor trick, our guide showed us how to get a close-up shot. Taking one of the cameras, he pressed it against the postcard and took a picture. He then showed us the result: the display window on the camera showed what appeared to be an extraordinary shot of a wild wolf striding across the snow and headed for the tree line. The effect was very nearly orgasmic. There were hoots of surprise, cries of "do it again" and a flurry of people running to the store to buy postcards to replicate the procedure.

The embrace of the hyperreal here is, of course, hard to ignore. The fact that photographs taken for the postcards were likely shot in other captive facilities did not seem to be in any way problematic. Instead, they were celebrated as something to show the folks left at home—a novelty. Neither wild nor free, these wolves could be collected, another indelible image to add to the catalogue of the trip. Their image, not seen but captured nonetheless through the lens of the camera, seems to bolster the guide's previous comment: "If they don't see it in their eyepiece it doesn't exist." The converse also seems to be true: if they see it in their eyepiece it must, in fact, exist. So, the best pictures from this jaunt through Grand Teton and Yellowstone, the ones that will receive the most attention and praise when travelers return home, are not those that capture (or likely fail to) the real lives of wolves or bison or bears, but rather pictures of postcards. And this, I think, describes one of the key paradoxes of this tour.

Day 5: Lamar Valley—Global Vision, Local Impacts

For the next two days, we stayed overnight in a town called Cooke City, Montana, population 140. Cooke City is located just three miles from the gate to Yellowstone and is part of the spectacular Beartooth Range, making it both a remarkable setting and a useful base to make forays into Yellowstone. This hamlet seems to function almost wholly at the behest of the tourist industry. Home to a disproportionate number of restaurants and hotels for its size, Cooke City is one of the gate communities whose existence is supported through the mass migration of tourists to Yellowstone in the summer months.

But, of course, this was not always Cooke City's economic base. For one hundred years, the fortunes of this town were tied to the copper, gold, and silver ore that lurked below the surface of the surrounding mountains. In the late 1800s, Cooke City, like many parts of the West, was born through the gold rush. This was doubly unfortunate for the area's original inhabitants—the Crow—who, having already been forced onto a reservation, now found the area that had been allotted to them squarely in the middle of the latest land grab. The Cooke City Chamber of Commerce's version of these events in their online history is both vague and sanitized. In reference to the second removal of Native Americans to make way for resource extraction, the Web site notes only that "In April of 1882 the reservation boundaries were released and the mountains were opened up to the awaiting prospectors" (Cooke City Chamber of Commerce n.d.). With the annexation of this territory, the "New World Mining District" was formed, which would later become Cooke City. Mining became and remained the raison d'être of the town until very recently. In the 1990s the economic base of this tiny town became unfixed, with debate between mining or tourism, "extractive" or "attractive" capital to use Timothy Luke's (2003) words, featuring as the central divide. This debate seemed settled when Noranda Inc., a Canadian mining company, bought the rights to mine this area as part of the modern gold rush. However, in 1996, after public outcry and presidential action in the name of national nature, Noranda agreed to a federal land transfer that amounted to approximately $65 million, although they had only paid $135 for the land under the federal 1872 Mining Act, which stipulated that lands could be sold for $5 per acre (Brooke 1996). This decision marked the transformation to a service rather than resource economy, and recreational tourism has now become the village's primary (pre)occupation. Indeed, of the ninety people in the Cooke City/ Silver Gate census area eligible for employment in 2006, sixty worked in the "arts, entertainment, recreation, accommodation and food services" industries, while only two people were employed in "agriculture, forestry, fishing and hunting, and mining" sectors (U.S. Census Bureau). This move to the service sector is tied to what some locals have termed the "yuppification" or "Californication" of Cooke City (Fishman 2010). Nature remains central to Cooke City's existence, but its role has shifted from productive to consumptive (ibid.).

Cooke City's main street provides a visual representation of the town's economic reconstruction. A simulacrum of the "Wild West," it sports a saloon, a trading post, kitschy storefronts, rustic motels, and wooden verandahs. While presumably bustling during Yellowstone's high season between June and August, in October it appears haunted, a ghostly and unpeopled reminder of a past that never was. One expects to catch sight of the clichéd tumbleweed making its way down a dusty street. But there are people here. While many have left for the season, some remain, and the tour is able to find a French restaurant and a gift shop to purchase yet more souvenirs to mark the journey. In this vein, the guide took us to visit a local wildlife photographer who sells his work to Yellowstone tourists. Indeed, the gallery seems to be an attraction on many trips of this ilk, and the photographer plays his part admirably, oscillating between sophisticated nature expert and down-home rugged individualist. I do not mean to suggest this is phony. But he does engage in a performance, and one that plays well with a sympathetic crowd. After a slide show of his remarkable images, we were invited to purchase prints, which many did. So not only do participants take photographs, but they have access to professional images to define their trip.

This act of consumption represents not only the penultimate expression of the visual grammar of nature explored above—by which I mean nature captured and rendered forever observable by the lens of an expert photographer—but also points to one of ecotourism's promised virtues: the opportunity to contribute economically to communities that chose to "save" their nature rather than "exploit" it. Through the purchase of wildlife photography, staying at the local inn, or eating in the French restaurant, ecotourists can feel as though they have not contributed to the tourist machine of, say, the cruise industry, but have instead had an experience of the wild while simultaneously ensuring its continued existence. Consumption saves the day.

But how true is ecotourism's claim? To be sure, tourist dollars remained in Cooke City. In our interview, the guide stressed this facet of the tour, arguing that this influx of money allows wildlife to thrive. Citing the example of the reintroduction of wolves to Yellowstone, he indicated that the money brought in by wolf-watchers has provided the economic justification to preserve nature rather than turning once again to resource. However, one wonders how much the bulk of the trip's proceeds, the over U.S. $2,000 spent on the ecotour, remained not in the local communities but with the ecotour operator's head office, several hundred miles away. In exchange for a turn to the aesthetic, the people of Cooke City, with a median household income of just $25,000 in 2000, are rewarded with contingent, seasonal, and often low-paid work of the tourist industry rather than the considerably more lucrative though possibly dangerous work of mining (1999 dollars, U.S. Census Bureau 2006). For example, the mean annual income for someone working in food preparation or service in Montana is $19,190, and in retail sales one can expect to earn an income of $28,470; for farming, fishing, and forestry the annual income is $30,340; for service unit operators in oil, gas, and mining, one can expect to make $54,860; and mining engineers come in at $71,390 (U.S. Bureau of Labor Statistics 2009). Indeed, in the Cooke City/Silver Gate census area, 14 of its 140 working inhabitants earned less than $10,000 per annum, with the bulk of the working population—30 people—earning between $10,000 and $49,000 annually (U.S. Census Bureau 2006). Interestingly, although some have noted that these jobs also tend to be racialized and gendered as well as classed, Cooke City appears to be something of an exception. In terms of race, this could be because, of the 140 residents of the census area, 137 are white (ibid.). However, gender is also equally split here, with 21 men and 16 women in service occupations like food preparation. In any case, it seems that ecotourism in Cooke City represents, no doubt, to some degree a mutually beneficial relationship among the tourists, outfits like the ecotour operator, and those employed in the industry. However, like most relationships of profit and power, not all players in this game are equal.

As explored above, Catronia Sandilands (2003) has suggested in a different context that the move from a resource-based economy to one that is preserved for recreation is not as unproblematic as it might appear. She asserts that by positioning places like Cooke City as natural, they are all the more inserted into the fray of global capitalism. Put another way, the move from extractive to aesthetic capital remains firmly within the constellation of capitalist enterprise, all the while maintaining its distance from the distasteful business of business. Much like Disney's Environmentality Program, what such gestures do is sidestep any kind of real critique of the structures and processes that have generated environmental devastation and imagine that solutions can be found within the very systems that privilege efficiency, economic rationality, and profit.

Day 6: Wolf Watching in the Lamar Valley

After five days of searching, standing in the cold, looking through scopes, and waiting, the final day of this trip provided the reward that all the ecotravelers were waiting for: a prolonged opportunity to watch the behavior of wolves in the "wild." While the bison had been initially captivating, their ubiquity quickly bred familiarity, and not even the most diehard among us continued to take photos of them after the fifth day. We had seen various other animals—elk, prong-horned sheep, deer, moose, ravens, eagles, and bears—but none really compared to the experience of watching wolves on the last day of the trip. We woke early that morning and headed out to the Lamar Valley before dawn. There we encountered a wolf kill. There were two wolf packs at the scene: the Druid Peak pack, which our guide speculated were the authors of the elk slaughter, and the Slough Creek pack, whose stronger numbers allowed them to take the carcass for themselves. But these packs were also not alone. Coyotes began to patrol a boundary around the kill, seeking to avail themselves of the bounty that had been provided. A grizzly bear also arrived, feasting on the elk corpse for one hour before, either sated or driven off by the constant nipping of the wolves, it wandered into the trees. We remained for approximately two and a half hours, taking turns looking through the scope using hushed and reverent voices to express our awe at the scene that seemed orchestrated as the apex of our trip to see the wildlife of Yellowstone and Grand Teton National Parks.

Of course, wolves were not always so venerated, at least not by the European colonizers of North America. The settlers brought with them biblical images of wolves in sheep's clothing, Brothers' Grimm and Aesop with their tales of big bad wolves, and horrific stories of marauding werewolves (Pluskowski 2006), all of which informed the mass extermination of wolves on the East Coast of the United States. As the colonial machine pressed further west, wolves were routinely cast as voracious beasts that slaughtered livestock at whim, making their so-called depredations legendary in the constantly shifting frontiers of an emerging nation. Equally legendary was the manner in which settlers dispatched their foe, with an efficiency and bloodthirstiness that some have named a "pogrom" (Lopez 1978). Wolves, much like Indigenous people, were cast in the role of anachronism and hopelessly out of place: "Civilization has the animals in its teeth. Cruel or humane, the manner of death mattered little, for wolves were already dead. Hopelessly out of place in rangelands dedicated to the growing of livestock, their extinction was unavoidable" (Coleman 2004, 207). "Progress" dictated the extermination of the wolf, and thus bounties were enacted to enable the opening up of the West, to protect valuable livestock and to consolidate land and territory to further the colonial agenda.

This view of wolves persisted through the formation of the national parks and into the middle of the twentieth century. The wolf had been resolutely judged an undesirable animal in need of management (read: extirpation). To ranchers, the wolf represented a threat to their property and livelihoods. To hunters, the animal's predatory skill provided too much competition. Even for those whose express wish was to see wildlife, the wolf's reputation as a bloodthirsty killer made it an unwelcome actor in western parks. Of course, this reputation could more easily be ascribed to those who dispatched *Canis lupus* with remarkable callousness and barbarity. However, this kind of critique was unsayable, particularly in the first national park, where animals could only be "useful, aesthetically pleasing or amusing" to find their place "in the 'outdoor zoo' of Yellowstone" (Jones 2002, 28). This meant that wolves were almost completely gone from the Yellowstone ecosystem by 1926.

Even still, wolves survived. Their persistence in spite of trappers, hunters, ranchers, bounties, strychnine traps, and predator control units has meant that wolves have lasted to encounter a new time, where they are viewed as the essence of wildness almost obliterated by the thoughtless actions of humans. Here we are reminded of Rosaldo's (1989) imperial nostalgia: we mourn that which we have destroyed. And so, wolves, along with bears, bison, and mountain lions have emerged as charismatic megafauna: to save them might mean we can save ourselves.

This romanticization was combined with a more secular impulse in the effort to recuperate the wolf. The birth of scientific notions of ecosystems, along with the naming of the gray wolf as critically threatened under the 1973 Endangered Species Act, have also worked to change the animal's fortunes. Ecologists began to reframe their absence as a story of imbalance—wolves were seen as the "missing link" in Greater Yellowstone ecosystem (Jones 2002). Indeed, in a short film run on a loop at the Albright Visitor's Center and Museum in Mammoth (screened before our observation of the wolves), the narrator indicated, "You can't have wilderness without them." At the same time, the Endangered Species Act mandated the reintroduction of experimental populations of species where their numbers had been decimated. Wolves found themselves on this list, and their reintroduction was scheduled for 1987. Of course, this move was actively contested: many ranchers, farmers, and townspeople still felt the threat of *Canis lupus.* Much like the 1988 fires, scientists were pitted against locals in the battle to restore/restory the Yellowstone ecosystem. However, the weight of scientific consensus, federal will, and the intensification of Yellowstone's insertion into the global market for nature images and experiences worked to mute the voices of those rooted in a different understanding of nature. The coming together of the romantic and the scientific sensibilities worked to justify governmental regulation to re-engineer ecosystem balance.

So began "Operation Wolfstock," the project to restore Canadian gray wolves into Yellowstone. In 1995 fourteen wolves were transplanted from Alberta to pens in Yellowstone to "acclimatize" for ten weeks in their new home. The NPS was concerned that their quick release might cause the wolves to attempt to cross the Rockies and find their way back to their original packs. They need not have worried. The new wolf population, finding itself in an ecosystem with few top predators, grew rapidly. Released from their small enclosure to a much larger, though no less real one, the wolves of Yellowstone have fared well, gaining celebrity status. The park now hosts 96–98 wolves in 14 packs and attracts approximately 20,000 people each year who engage in a variety of wolf-watching activities (Smith et al. 2007; Yellowstone National Park 2009). By all accounts bringing the wolf back to Yellowstone represents a successful species reintroduction.

And yet, something worries this neat picture, gnawing at its precise margins. Coleman (2004, 227) contends that the reintroduction of wolves to Yellowstone is as much about culture as it is about nature, and perhaps more about ideology than biology. Indeed, he asserts that restoring this ecosystem's keystone predator is a way to "safeguard . . . wilderness fantasies" rather than save a species from extinction. I actually think it is a bit of both. More precisely, I think that this act of reintroduction is a biopolitical project, not only for the animal it purports to manage but also for the humans who come to gaze upon it. The return of wolves to Yellowstone works on a number of biopolitical registers. First, on a meta-level, this reintroduction speaks to the possibility of remaking nature; a pristine wilderness, much mourned though necessarily sacrificed for the sake of progress, could be recreated. Reminiscent of the *Jurassic Park* (1993, 1997, 2001) film series, humans can actually engineer ecosystems and render them back to their supposed original state. Of course, there is no tinkering with DNA trapped in fossils in Yellowstone; instead, Canadian wolves were substituted for American (of course, the wolves themselves make no such nationalist distinctions). However, the psychic impact is quite similar. If we can enliven habitats that were once decimated, or at least imbalanced, this affords the possibility of an intellectual leap. It allows not only for the imagining of premodern "virgin" wilderness but also that this wilderness can be returned to a previous form through the judicious application of knowledge around the health and well-being of ecosystems. In Yellowstone, there was a perceived lack, which was remedied through the scientific and technological mastery over nature. The management of nature in this respect can authorize its destruction of nature, knowing that it can, once again, be restored.

If ecosystems are regulated here, so too are the animals and humans who sometimes inhabit them. The wolves of Yellowstone have been thoroughly monitored, tracked, and collared. Their kills are documented, their conflicts observed, their reproduction assessed, their dens visited, and their DNA sampled (Smith et al. 2007). Some are equipped with satellite collars that e-mail the precise location of wolf packs to staff at the Yellowstone Wolf Project. In fact, we only spotted the wolves because our guide was in contact with staff from the NPS and the Yellowstone Wolf Project who track the wolves via radio telemetry. These wolves straddle the boundary between wild and tame, captive and free. In some ways, they are subject to management strategies similar to those one finds in zoos. In others, scientists and NPS staff take a more hands-off approach, allowing the ecosystem to regulate itself, all the while taking assiduous notes. However, each approach is part of the same system of regulation, and each produces biopolitical information. In the first sense, staff managed the wolf populations to diagnose illness, assess eating habits, chart conflict, map DNA, and, generally speaking, ensure well-being. But even as scientists allow the ecosystem to "manage itself," they still study the wolves so that the population can be known and recorded—all of its facets documented for scientific pursuit, scholarly interest, or perhaps posterity.

Simultaneously, human visitors are brought into this regime, as their affective desires to experience these now mythic animals are mobilized and managed and a wolf industry is born. The people who go and see wolves, bison, and bears are drawn into this fictive endeavor where nature remade and regulated is imagined as pristine, wild, and free. Jane Desmond (1999) has cogently noted that what makes an ecotour different from other sorts of tourism is its chimerical "authenticity": it carries the appellation "natural," which necessarily means it lacks contrivance. Whereas the exhibits of the American Museum of Natural History and the simulations of Disney deploy artifice to speak truth, ecotours operate in and through the notion that what is seen is the genuine behavior of animals in the wild. In doing so, the ecotour denies the whole range of nature–cultures that make up these places of wilderness consumption. If history, culture, and power are erased from places such as Yellowstone or Grand Teton (or the Galapagos Islands and African safaris, for that matter) through violent acts of discursive and material purification, then these spaces become not lived realities—either by human or nonhuman nature—but rather museum pieces, to be visited, gazed upon, remarked about, photographed, and left eventually behind.

Conclusion: Leaving Wonderland

After watching the wolves, coyotes, and a grizzly battle on the floor of the Lamar Valley for two hours, it was time to return to Jackson for the denouement of our trip. The van ride proved long, because we were near the border between Wyoming and Montana, providing the ecotourists with time to digest what we had experienced throughout the week. It seemed as though my fellow travelers, imbued with both a feeling of exhilaration and a sense of responsibility, needed to find ways to make sense of this trip. How can we bring these experiences back home? How can others learn, as we have, about the importance of nature? How can what we have seen be shared with everyone? These are the questions that informed the discussion on the van ride back to Jackson. There was a general feeling that the problem was numbers: the critical mass necessary to support preservation is missing. By way of emphasis, one participant indicated, "Everyone, especially school children, should come on these excursions." Sage and somewhat melancholy nods accompanied this exhortation. No disagreement here. Just lament that our journey was over and that others have not experienced nature as we did.

The above declaration points to the ways that ecotours can govern experiences of nature. Most obviously, the assertion that all people should go on ecotours is blind to the class dynamics at the core of this kind of nature consumption. There was no discussion that this sort of experience is not available to everyone when, at the low end, it costs three thousand dollars a shot. Moreover, that differently situated people might understand this ecotour—and its selective preservation, rupture between nature and culture, and fetishization of vision—differently does not appear to hold any sway. Rather, there is a normal way of encountering nature: away from home, singular in its grandeur, aesthetic in composition, and devoid of humans. There is no sense that what is seen is transformed by our gaze, or that we operate as an assemblage of lens/eye/animal in the scene that is imagined as nature unfolding as it would without our presence. Rather, there is the view that the nature we have come to see is just that: natural. It is this kind of nature that all people should experience and understand in the same way.

The statement that all people should experience nature through largely watching it from within or near to a van is lent coherence by the discourses that circulate about how to save global nature, as discussed in chapters 1, 2, and 4. There can be no discussion of what it might really take to reha- bilitate ecosystems like Yellowstone, all the while accepting that it is nei- ther pristine nor wilderness. Moreover, like parts of the Hall of Biodiversity and more expressly Disney's Animal Kingdom, the environmentalism on offer here provides no critique of the consumer culture, corporate practice, global environmental injustice, the violence of preservation, or animal incar- ceration. To do so would be to critique the very idea of ecotourism and the entitlement of travelers who believe they have a right to see, know, and collect distant natures in locales other than their own.

But I want to suggest a little bit more than this. What I want to say is that there is a particular technology of vision that operates through the ecotour, which works to govern how the nonhuman is consumed. Of course, scholars of both tourism and ecotourism have picked up on Foucault's notion of the gaze and panoptics, as most fully elaborated in *The Birth of the Clinic, The Order of Things,* and *Discipline and Punish.* As outlined ear- lier in this chapter, such theorists have paid important attention to the ways that the tourist gaze circumscribes space, delineates view, and seeks to know and represent the essence of a place. The notion of the gaze is a central facet of this ecotour, where vision (and its attendant affective dimensions) becomes the primary vehicle for nature's apprehension. Fou- cault articulates in "The Eye of Power," that the gaze works through the panoptic vision of Bentham's prison, where surveillance is exercised such that individuals become their own "overseers" (1980, 155), limiting their vision to that which is beautiful, majestic, and wild.

But what if the eye isn't always so masterful or disciplined? And what if we reconfigure the gaze as more of an encounter? In the first chapter of *The Order of Things,* Foucault uses Diego Velazquez's 1656 painting *Las Meninas* as a heuristic, working with it to think about how vision, subject, and object are organized. *Las Meninas* is a painting of a painting, where the distinction between spectator and those looked upon seems to fall away: "No gaze is stable, or rather, in the neutral furrow of the gaze piercing at a right angle through the canvas, subject and object, the spectator and the model, reverse their roles to infinity" (Foucault 2004, 5). It seems, in some sense, that this painting looks back. Here Foucault speaks to the complex character of the gaze; the hard boundaries between those that look and those that are looked upon is more slippery than they might first appear. But there is privilege in this looking, and it raises the questions about beauty, subjectivity, and power in interesting ways.

So, I want to complicate this notion of the gaze somewhat. In the final analysis, the ecotour acted as a particular kind of school, a vehicle, like the others in this book, for the production of a particular way of understanding the biophysical world. It did not generate scientific knowledge, like the museum or even the Animal Kingdom. Rather it worked to dispense the truths of science, and to demonstrate how these might be applied to understanding nature and living properly through an encounter with the putative wild. But the exploration of the biotic features of ecosystems, while important to the guides, was not what the tour ended up trafficking in. Instead, what the tourists take away is an affective understanding of the importance of natural beauty. So, the ecotour was like a school, or perhaps a kind of vision machine. Indeed, it served as a kind of intellectual factory for the production of a green governmentality focused on aesthetics.

The entire tour was framed through the eye of the ecotourist. Indeed, each aspect of the ecotour supported this focus on aesthetic beauty. Visits to Old Faithful, the most photographed barn in the United States, and Cooke City offered opportunities to experience nostalgic landscapes, freighted with notions of nature and nation. But more important were the opportunities to view animals, the hidden treasures of this parcel of the United States. Wolves, most particularly, but also bison, bears, elk, and moose were the charismatic megafauna that made the trip possible. Because, of course, one could quite easily visit either Yellowstone or Grand Teton on one's own and see the majestic landscapes that have been so often reproduced. It was the access to these animals that the travelers paid for and which the guide provided in spades, largely because species like the wolves are so regulated in national parks. Landscape, animals, and guide expertise worked in tandem to produce a visual grammar of "pristine" national natures. Each day of the ecotour highlights the majesty of this region, endangered animals that must be seen, and how it all may be represented in the assemblage of lens/eye/animal. As such, there is a normal way of knowing the wild as largely unpeopled, awe-inducing, exceptional, and always aesthetic in composition.

And all these landscapes and animals were captured through the technology of photography. Through photography, the tourist gaze worked to fix places like Old Faithful or the Teton Range, as well as animals like bison, wolves, and bears. This fixing generates meanings that are removed from their historical, economic, cultural, biophysical, and political surround. Through this ecotour, nature was both incarcerated and made legible through display, a move that regulates that which can be seen and known. In the selection of what was important to see, experience, and understand, this ecotour delimits that which counts as nature and hence, that which does not; vision and experience were organized so nature can be understood in particular ways (as romantic wilderness, scientized ecosystems, or aesthetic experience) by particular kinds of people (environmental subjects who are often white, moneyed, and seeking the animal "other"). In doing so, ecotourism works to provide a simpler kind of environmentalism, one in which consumption equals action, and traveling to sites of "wild" nature makes one greener. And so, this brand of green governmentality cannot operate without commodification. Ecotourism provides yet another space where nature is transformed into commodity. In this move, nature, and how it is understood, can only be changed by the encounter. By selling the experience of nature, marketing it as a genuine encounter with pristine wilderness, this ecotour worked to not only produce profit, but also power.

The impacts of this way of understanding nature have weight. Of course, the ecotour stands alone neither in its construction of both vision as the primary sense to understand the wild, nor in its emphasis on particular kinds of splendor in nature. Indeed, it connects to a long history of wilderness as sublime. But it does offer an encounter with this production of legibility that, for example, the photographs of *National Geographic* or the stunning cinematography of *Planet Earth* does not. And the particular legibility offered through the ecotour forecloses the very possibility that nonhuman nature is not just beautiful, wild, and photographic, standing in radical alterity to humans, but rather is an assemblage of unruly combinations of both human and nonhuman, nature and culture. Indeed, what such distinctions do is reify such categories, which may not be the most useful and certainly are not the most interesting ones to think about our world. As seen on the trip, this aesthetic green governmentality allowed for a host of divisions to be made; for example, that some animals are wild (and thus imbued with markers of freedom, nobility, and beauty) while others are tame and banal, or at least unthought of as part of the same bio-physical world. Moreover, the animals that are deemed wild are increasingly brought under a kind of regulation, so that, ironically, they can continue to be viewed as such. And other landscapes become, if not illegible, then irrelevant as sites of nature.

This focus on aesthetics presumes a distance, but also an invisibility of the presence of those that look. And yet, if we return to *Las Meninas,* I think there is something interesting to be said about seeing and how it can work as a technology of green governmentality. Because, like *Las Meninas,* the animals and landscapes gazed upon on this ecotour are not passive objects; they *look back.* Instead of a one-sided relationship, what is on offer through the ecotour is an encounter, after which neither human nor non-human are the same. Rather, there is a complex web of vision that governs, but can never be completely managed. Of course, this relationship is asymmetrical. The wolves, for example, that inhabit the Lamar Valley are in some sense subject to the gaze of the ecotourists and the NPS staff who seek to access and understand their lives. So too are the elk, bison, and bears. But the nonhumans in this regulated space also e/affect those who look upon them. And they go about their lives in ways that, although interesting to their human watchers, also work for themselves. So, while I want to assert that there is a kind of aesthetic green governmentality here, I contend that it is more complicated than it might appear, chock full of agency, slippage, and relationships of power, where subject and object can often be confused. In some sense, while demonstrating the power of vision, I think the ecotour offers an opening to think about the relationship between the human and nonhuman in different ways—potentially a way out of governmentality.

CHAPTER 4

Science and Storytelling:
Al Gore and the Climate Debate

I N THE LAST FEW YEARS, it seemed there was the potential for consensus, or at least uneasy quiet, in the climate change debate. After decades of vacillation, obfuscation, and the production of uncertainty around this issue, particularly in the United States, the mounting scientific evidence appeared to have opened a political window of sorts, bending the weight of public and political opinion toward the understanding that climate change was an issue that required action. Even staunch "deniers" like George W. Bush were required to recognize anthropogenic global warming as fact. Indeed, both U.S. presidential candidates in 2008 recognized climate change as a reality. Clean energy, the green economy, and the risks associated with oil dependence peppered the stump speeches of Barack Obama and John McCain alike. To be sure, there were still those who maintained that climate change was a fiction, but they seemed, to some degree, bound within the confines of right-leaning media and quack science. It appeared as though some kind of action was imminent, and the debate would focus on the appropriate methods to both mitigate and adapt to the climate crisis.

However, the recent events of what was dubbed "Climategate" has shattered this rather tenuous calm and opened a space for continued dissent. In a sensational spectacle worked through the world's media outlets, Climategate was the publication of hacked e-mails among various scientists at the University of East Anglia's Hadley Climatic Research Unit (CRU). That correspondence was, for some, evidence of scientific malfeasance. At the forefront of the ensuing maelstrom was Dr. Phil Jones, Director of the CRU, who, along with his colleagues was accused of suppressing and manipulating data that contradicted global warming "orthodoxy," as well as tainting the peer-review process to keep out those views that did not correspond to the view that anthropogenic climate change was a proven fact. Conveniently, these e-mails and related documents were leaked to the press just before the Copenhagen Conference of Parties on climate change in November 2009. In the flurry of media excitement following the release of these e-mails, some suggested that this revealed that global warming is a hoax, with the British newspaper the *Telegraph* stating that Climategate represents "the worst scientific scandal of our generation" (Booker 2009). Others in the media were less radical in their assessment; Dr. Peter Keleman (2009) noted in *Popular Mechanics* that while the e-mails may be damaging, in no way do they dismantle the scientific consensus around anthropogenic climate change. Soon after Climategate, Al Gore weighed in, writing an op-ed for the *New York Times* that elaborated this position, reiterating that although mistakes are often made in scientific research, "the overwhelming consensus on global warming remains unchanged" (2010).

The degree of separation between those who argue that anthropogenic climate change is a reality versus those who assert that it is a scientific conspiracy is no real surprise. Regardless of the outcome of an independent tribunal that has largely cleared the CRU of wrongdoing, the impact of Climategate has been to destabilize the putative climate consensus, lending some legitimacy to those who suggest that climate change science has been faked. Keleman suggests that "[p]erhaps the most worrisome part of this incident is that it could easily leave the public wondering about the science of human-induced global warming" (2009). It could and by some accounts already has led to a certain degree of public skepticism. For example, a recent survey was conducted by researchers at Yale and George Mason Universities that compared answers to questions about concern for and evidence of climate change in 2008 and 2010. The interviews were conducted from December 24, 2009, to January 3, 2010, just after both Climategate and the Copenhagen summit. The survey results released thus far showed that there is a marked increase in uncertainty about climate change. For example, in 2008, 71 percent of respondents indicated that they believed climate change was happening; in 2010, that number shrank to 57 percent. On the question of whether most scientists think global warming is happening, the number reduced from 47 percent to 34 percent (Leiserowitz, Maibach, and Roser-Renouf 2010). While these changes cannot solely be placed on the revelations of Climategate as shaped by the media, it certainly has helped to shift the terrain of debate.

What Climategate demonstrates is that media *matters* in the construction of environmental issues. Newscasters, journalists, pundits, and celebrities work to shape public perception of the issue of climate change, and hence what strategies might be employed (or not) to mitigate it. "Climategate" is one salvo in the larger battle between climate change scientists and skeptics, a conflict that is most often fought through multiple sources of media. Indeed, most people come to understand climate change not through the reports of the IPCC but rather through the ways in which this information is filtered through news sources, celebrities, public figures, politicians, environmental groups, and so on. This means your view of climate change, depending on whether you watch Fox News or listen to NPR, read *USA Today* or the *Guardian,* is shaped by a particular branding of the climate debate. Moreover, what this attention to media also suggests is the importance of trustworthiness; the character and actions of the subject who "speaks the truth" of climate change are central to media representations of it. Who can claim credibility and access media sources are a central part of the climate discourse.

This chapter delves into the construction of climate change discourse by examining the personage and work of perhaps the most compelling media figure in the climate debate: Al Gore. Carrying the monikers "Al Bore," "The Ozone Man," and my personal favorite, "The Goreacle," Al Gore has been part of the American political consciousness since he was a child, son of prominent Tennessee senator Albert Gore Sr. His biography is important, not only to elaborate the details of how he became the most famous personality in the climate change debate, but also because the arc of *An Inconvenient Truth* also narrates the story of Al Gore, weaving together his life as inseparable from the climate crisis he works to forestall.

Making the Goreacle

A flurry of biographies came out just as Gore ran for the presidency in 2000. They described his early upbringing and the events that led him to this race. Two in particular are noteworthy: Bill Turque's (2000) *Inventing Al Gore* and Alexander Cockburn and Jeffrey St. Clair's (2000) *Al Gore: A User's Manual.* There are some similarities between the books as might be expected, given that they cover much of the same historical ground, although the tone of each book is quite different. Bill Turque, a journalist with the *Washington Post,* suggests that Gore's story is characterized by a commitment to his political ideals and a depth of policy knowledge, but also a tedious attention to detail, a difficulty connecting to voters or fellow politicians, and a tendency toward self-interest and calculation. Cockburn and St. Clair, founders of the radical newsletter and Web site *CounterPunch,* offer a more politically charged analysis. Indeed, in the opening page of

their treatise, they argue that Gore "distills in his single person the disre-
pair of liberalism in America today, and almost every unalluring feature of
the Democratic Party" (2000, 1). I rely on both of these books, as well as
numerous online biographies, to briefly narrate the making of the Goreacle.

Gore has been excoriated in both the mainstream media and the popular imagination for supposed exaggerations of fact that, many have suggested, have led him into the realm of fiction. For example, much was
made of a 1999 interview with Wolf Blitzer on CNN where Gore asserted
that he took a leadership role in the expansion of the Internet. While he
was instrumental in pushing through the High Performance Computer and
Communication Act of 1991, which greatly expanded the Internet, this
statement was taken up in the media as an assertion that he invented the
Internet, something which would be a rather large aggrandizement given
its creation by the military in 1969. However, the suggestions of embellishments don't end there. It has been argued by various critics in the press
and beyond that Gore claimed credit for adding to Hubert Humphrey's
1968 Democratic nomination acceptance speech, which proved to be false.
Others intimated Gore claimed that he and his now estranged wife, Tip-
per, served as the model for the film *Love Story,* a comment disputed by
the author of the screenplay as an overstatement (Turque 2000, 22–23).
Further, when Gore worked as a reporter with the *Nashville Tennessean,* he
contended that one of his pieces on municipal corruption "put a lot of
people in jail," which also proved to be overstated. In each of these cases,
as Turque describes, there is an element of truth to the story that is amplified
to heighten Gore's prestige or accomplishments, although often it is clear
that the exaggerations could be less attributed to Gore and more often to
his critics. Turque suggests this penchant for aggrandizing elaboration was
encouraged by his parents, who saw Gore as a presidential candidate almost
from his birth. Born in 1948 as the son of Albert Gore Sr., senator from
Tennessee, and his wife, Pauline, a lawyer and political powerhouse, Al Gore
was prepped for his eventual bid for the White House. The Gore dynasty
encouraged such embellishments as part of the narration of the life of a
future president, and Gore was a boy who wanted to please and exceed the
high expectations of his overachieving family. Cockburn and St. Clair
describe him as "a display item to his parents' political associates and the
press" (2000, 12), always the perfect accessory to the political campaign
and one that was always mindful of his father's political image. Gore's early
life involved shuttling between the Fairfax Hotel in Washington, D.C., and

the Gore tobacco farm just outside of Carthage, Tennessee. He attended St. Albans School in Washington, D.C., a breeding ground for the next generation of political elite. He then went on to Harvard, eventually settling on a major in Government Studies, meeting people like Roger Revelle and Martin Peretz, who would be influential in his life. After much debate, and in line with his sense of political calculation, Gore enlisted in the army after finishing his degree in 1969, although at least in Turque's account, he was against the war in Vietnam. This is where political expediency came to the fore again; he enlisted to both support his father's reelection campaign and to secure his own ability to run for office in the future. He worked as a war reporter, though he was carefully shielded from any real violence by his political connections (Turque 2000, 82–89). He returned to the United States in 1971 to go to Vanderbilt Divinity School and begin a career as a journalist with the *Nashville Tennessean,* a well-known and politically connected newspaper.

His political life began more formally in 1970 with his first run for Congress, a campaign that he lost. He did win as representative in 1976, and later went on to claim his father's senatorial seat in 1984. Gore was guided throughout his political career, according to both biographies, by strategic opportunism and poll-politics, working on issues like arms control, tobacco labeling, the environment, and governmental oversight. But, at least according to Turque, he was dogged in his examination of these issues and fully committed to enacting strong legislative packages. In 1988 Gore made a failed bid for the Democratic nomination for president. He bowed out of the race, but not before, it has been suggested, he raised the specter of accusing Dukakis of being soft on crime. Cockburn and St. Clair contend, although this has been hotly contested, that he was the author of what became the Willie Horton debate, a narrative that the Republicans adopted and ran with after Gore's concession, and one that irreparably damaged Dukakis as the Democratic nominee (2000, 123–34). He continued to serve as a senator, and between 1989 and 1991 he penned *Earth in the Balance,* published in 1992. It was also during this time that Gore, by his own account, began to give the slideshow on which *An Inconvenient Truth* is based. Part scientific tome, part quasi-spiritual reflection, part autobiography, *Earth in the Balance* is Gore's effort to elaborate his position on what he saw as an impending environmental crisis, brought on by a disconnection from nature and an overconsumption of resources—in essence, a lack of balance. Gore asserts the book was precipitated in large part by the near death of his son, Albert III, hit by a car when leaving a Baltimore Orioles baseball game in April 1989.[1] This event, combined with his loss in the Democratic primaries, offered Gore a new lens with which to view the world: "This life change has caused me to be increasingly impatient with the status quo, with conventional wisdom, and with the lazy assumption that we can always muddle through. . . . We must all become partners in a bold effort to change the very foundation of our civilization" (Gore 1992, 15). The book covers much of the same ground that is abbreviated in the slideshow (down to the inclusion of the same anecdotes; for example, the kid who wonders if the continents fit together); it focuses on global warming, the history, science, and politics behind it, and the potential for universal catastrophe if human beings don't change our ways. And yet, the book carries a more strident tenor in many respects, criticizing free-market economics and government expediency, and asserting we inhabit a "dysfunctional civilization" that is "addicted to the consumption of the earth itself" (220). He offers six strategic goals in his new version of a global Marshall Plan: stabilize world population, develop environmentally appropriate technologies, establish a global economic system that takes ecological costs into account, establish international environmental agreements, coordinate a global education effort on environmental problems, and finally, create the social and political conditions necessary for these changes to be implemented (305–7, and further elaborated 307–60).

Nineteen ninety-two was an important year for Gore; not only did he publish his book that year, it also represents his election to the position of vice president in which he served from 1993 to 2001. The Clinton/Gore years were marked by a move to the right for the Democratic Party, and Gore was a key actor in this shift. Indeed, Cockburn and St. Clair (2000) name Gore as the most powerful vice president in history (this, of course, was before the Bush/Cheney administration). Although Gore pushed hard for the British Thermal Unit (BTU) tax as a rider on an economic policy bill soon after he entered office—a measure that would have taxed the burning of fossil fuels—the initiative failed to pass through the Senate Finance Committee (Turque 2000, 270–71). Despite this strong beginning, the remainder of the administration's environmental record was largely disappointing, especially for green groups who had imagined Gore as their champion. As both Cockburn and St. Clair and Turque outline, Clinton and Gore reneged on the pledge to implement higher fuel efficiency requirements on cars; broke their promise to increase grazing taxes in the American

West; focused more on free trade than environmental regulation, with particular impacts on forests; and failed to implement tough standards for strip mining, offshore drilling, and pesticide use. This might be due, in part, to the election of a Republican House and Senate in 1994, led by the vociferous Newt Gingrich and conservative Democrats in both the House and Senate. In *Earth in the Balance,* Gore contended that "perhaps most important, I have become very impatient with my own tendency to put a finger to the political winds and proceed cautiously" (1992, 15). But this approach seemed curiously absent when dealing with environmental issues. Instead, Gore focused on such issues as restructuring and streamlining government and a fairly disastrous foreign policy, which included interventions in Somalia, Haiti, and Bosnia.

And yet, Gore retained the image of a green politician, despite his record as vice president. He won the Democratic presidential nomination in 2000 and once again seemed willing to tone down his rhetoric to win votes. In the end, Al Gore won the popular vote, but the world was witness to the protracted battle for the actual presidency that hinged on the Florida recount to determine the electoral college vote. In December 2000, the Supreme Court decided to end the recount, and George W. Bush was declared president. Many felt that Gore was robbed of the election and rallied behind his potential candidacy in 2004 and 2008. But Gore had moved on to other endeavors, including the cofounding of Generation Investment Management in 2004, an equity firm that makes sustainability a key focus. Further, Gore became a partner in the green venture capital firm Kleiner Perkins Caufield and Byers, an association that some suggest has positioned Gore to become the world's first "carbon billionaire" (Broder 2009). In addition, he founded Current TV, a cable network that broadcasts both user content and produced shows from investigative journalism to movie reviews.

Of course, it is through the documentary *An Inconvenient Truth* (2006) that Gore has been brought both renewal as a political actor and a certain degree of veneration. Through the film and his attendant efforts to generate movement in the climate debate, Gore has generated a kind of significance and popularity that escaped him during his political career. Leaving aside his reputation for being wooden and embracing his distinction as a fact-oriented science wonk, Al Gore has emerged from his loss during the 2000 presidential election as a new kind of cultural force. Thus, Al Gore is at the forefront on a new politico-cultural category—the climate celebrity.

His mission is beyond that of politics or any American president; he has taken up the office of what Tim Luke wryly names "Planetary Proconsul," endowed with a mission to save the planet from ourselves (2008, 1813). Al Gore does not stand alone here; actor Leonardo DiCaprio, Virgin CEO Richard Branson, and environmental lawyer and activist Robert F. Kennedy Jr. can also lay claim to this title. In what Boykoff and Goodman have called "a new form of 'charismatic megafauna'" (2009, 339), celebrities are made into truth-telling global environmental citizens, herding (or attempting to herd) the rest of us along the path to green salvation through spectacle, technology, and good, old-fashioned public relations. But he is one of the most successful and credible of these climate change celebrities, and his views carry a great deal of weight.

Gore has translated this celebrity into a number of initiatives, some of which will be covered in much further detail later in this chapter. The point is that although his carbon crusader status was generated through the popularity of *An Inconvenient Truth,* he has spun it off in a number of directions, namely a series of books, nonprofits like Repower America, the Alliance for Climate Protection, and the Climate Protection Action Fund, and through his green venture capitalist firm, Generation Investment. Thus, more than, say, Leonardo DiCaprio, Gore has transformed himself into a guru of climate politics, science, and policy, working as both the generator of knowledge around the climate crisis and the potential source for its solution. But all this really began with *An Inconvenient Truth,* and so that is where we must start.

Storying the Science of Climate Change

In cultural and economic terms, the impact of this film is hard to dispute. Released in 2006, the documentary immediately won audience favor and critical success. In 2007, it won an Academy Award for Best Documentary and Best Original Song, an "Eddie" (American Cinema Editors Award) for Best Edited Documentary Film, and the Critics Choice Award from the Broadcast Film Critics Association, and Melissa Etheridge's Oscar-winning song, "I Need to Wake Up," was nominated for a Grammy. These are but a few of its many accolades. In addition to its many awards, the film has also garnered $50 million worldwide, making it one of the highest grossing documentaries of all time. The critical praise and market appeal were likely a large part of Gore's shared Nobel Peace Prize (with the IPCC) for his efforts to educate the world about the perils of global climate change. And if you Google the film's title, which one should given that Al Gore acts as senior advisor to the corporation, you come up with more than 9,190,000 entries, depending on the day.[2] Thus, it is not a stretch to say that *An Inconvenient Truth* has cultural importance, a kind of currency with which political and moral lessons are made and told.

The lessons the film provides are not those from which I pretend to stand apart. It is not my intention to set up a straw man in Al Gore that I will bash to pieces with the blunt club of deconstruction. For me, in large part the film is about both *knowledge* and *affect*, the capacity to know, but also the capacity to feel. When I saw *An Inconvenient Truth* for the first time in 2006, I didn't really know what to expect. I had heard about the movie from friends, both those active in environmental causes as well as those who never envisioned themselves as environmentalists, that it was a powerful movie, perhaps even a wake-up call. Aware of the science of climate change (though to be sure, no expert), I was astounded by the images presented in the film and both frightened and saddened by the prospect of the global future that Gore suggests is a possibility without action on this issue. And I admired Gore, the forthrightness with which he demonstrated his argument, and simplicity with which he asked the viewer to work for a better world. So while the knowledge generated through Al Gore's intervention into the discourse of climate change was important, so too was the bodily response it aroused: anxiety in the pit of my stomach, shame for my participation in carbon colonialism, and hope for a better future. As such, I think it's a valuable, complex, and complicated film, which acts as a clarion call for a particular kind of environmental consciousness.

For those who haven't seen the documentary, *An Inconvenient Truth* is really the chronicle of Al Gore and his quest to promote awareness about climate change. There are, of course, many ways to craft the stories that any documentary tells. The documentary style that most closely describes the structure of *An Inconvenient Truth* is the expository documentary. This brand of documentary focuses on the construction of an argument through the arrangement of a persuasive set of facts that the filmmaker uses to educate the audience. In the case of *An Inconvenient Truth*, as well as many other documentary films, the argument is organized around the "problem/ solution structure" (Nichols 2001). Nichols suggests that the expository argument has five component parts: a compelling opening; a rehearsal of what is known and what remains in question; the assertion of a particular position on the facts; a rebuttal of counter-arguments; and the culmination of the argument into a suggestion for action (2001, 56). This structure could be mapped almost exactly onto *An Inconvenient Truth*. Thus, the goal of the expository documentary is *to educate:* to elaborate a particular problem, flesh out the contours of the debate, and provide a solution through which viewers can act on the issue as it has been presented to them.

The goal of education is achieved through two interwoven narratives of science and storytelling in *An Inconvenient Truth*. The scientific dimension is represented by the storying of Al Gore's slideshow, which by his own account he has given over a thousand times all over the world. Edited by cobbling together three instances of Gore's performance, the slideshow sketches out the argument that anthropogenic global warming is an urgent reality in need of action. Gore and director Davis Guggenheim make this argument through rhetoric and image, deploying Gore's considerable ability to explain the concepts and data of climate science. He charts the basic science of global warming, looks at temperature rise for the last six hundred thousand years, deploys visualizations of sea level rise, offers information about the potential collapse of the Antarctic and Greenland ice shelves, shows images of melting glaciers, and discusses the potential for positive feedback loops and abrupt climate change, all the while exploring the impacts that these environmental events will have on nonhuman and human life alike. Sometimes humorous but mostly shocking, Al Gore's scientific exegesis provides the building blocks to understand the global scientific consensus that anthropogenic climate change is both real and threatening.

But the arc of the film is not fulfilled solely through an attention to scientific matters of concern. In equal parts, the film charts Al Gore's biography—the personal and political struggles that have led him to be a champion, perhaps the most preeminent champion, of this cause. And so, alongside his time-lapse projection of sea level rise and animation of drowning polar bears, there is a complementary narrative: Al Gore's life growing up in both rural Tennessee and in the halls of power as the son of a senator, his early introduction to climate science at Harvard University with Dr. Roger Revelle, the near loss of his son, the death of his sister, his defeat in the 2000 election, his resurrection from political obscurity, and the finding of his political purpose. All of these elements provide a backstory that serves two purposes: to venerate Al Gore as hero and prophet, and to make the science he narrates seem all the more objective, legitimate, and chilling.

Instead of examining each frame of the film in detail (an impossible task that would read more like a screenplay than an analytic venture), I turn now to a few vignettes from this documentary to elaborate on how science, celebrity, biography, narrative, and power come together in *An Inconvenient Truth*.

Science, Knowledge, Power

Scene: Al Gore stands before a packed audience giving his global warming slideshow. He has already opened with his stock joke—"I'm Al Gore and I used to be the next president of the United States"—and has moved through the basic process of global warming as well as a short and humorous *Simpsons*-style animated clip (written by his daughter Kristin). The introduction is over; it's now time for the hard truth of scientific fact. The audience is witness to what the last thousand years have shown in terms of temperature rise and CO_2 concentrations. He uses this to debunk arguments around the Medieval Warm Period and to demonstrate that temperature has risen most sharply in the last few decades, along with CO_2 emissions. But Al Gore reserves his most striking image for the record in Antarctica, where this data can be tracked for the last 650,000 years. He tells us that this is the first time that anyone apart from a small group of scientists has seen this image. Gore stands dwarfed by this massive graph and we watch as red and blue lines streak across the black screen that render palpable what is business as usual in climate fluctuation, and what is decidedly abnormal. He alerts us that within the time frame of the graph, CO_2 has never gone above 300 parts per million. However, the impending catastrophe is made clear when Gore illustrates how far above this benchmark we've risen. In a humorous moment, and in Al Gore's typically folksy way, he climbs atop a hydraulic lift to reinforce his point. But the giggles soon fall away as the graph's red line, CO_2 concentrations, extend far above the scale of the chart, requiring an extra screen to display it. If fossil fuel consumption isn't slowed, this is what CO_2 concentrations will be in the next fifty years. He reminds us of his earlier comment, that the difference between a nice day and a mile of ice over our heads is but a short distance between the red and blue lines. What, then, will be the effects of this now yawning gap? Gore then tells us: "Ultimately this is not a political issue so much as a moral issue. If we allow that to happen, it is deeply unethical." And if its an ethical issue, then Gore sets himself up to tell us the right path forward to ameliorate our transgressions.

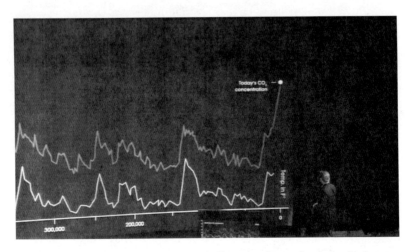

Taken from the film An Inconvenient Truth, *this image depicts the relationship between carbon dioxide and temperature over the last 650,000 years.*

What is interesting about this scene is that so many of its elements resonate throughout the movie. Gore employs scientific evidence as one of the key discourses to narrate the import of climate change. However, he does this through two interrelated mechanisms. The first is his usage of technologies of vision. Gore and director Davis Guggenheim are not simply giving their audience the latest analysis of the IPCC. Their task is rather to make the science breathe, to give it flesh, so that the audience can see and feel the weight of its predictions. This is done through the use of image, which not only makes the science more comprehensible and consequential to the viewer, but also serves the media requirements of a major motion picture to produce awe, fear, and drama. Hence, *An Inconvenient Truth* uses images, statistics, simulations, temperature predications, diagrams, and graphs—in short, technologies of vision and tools of intelligibility deployed through science—to map the state of the globe's nature and to give a compelling lesson of what happens when politicians and citizens do not listen to the wisdom of science. Thus, the film uses an already accepted discourse of "truth" to narrate the appropriate way to save nature.

But the visual representation of science alone cannot do all the work that *An Inconvenient Truth* requires. The facts, it seems, do not speak for themselves. Which leads us to our second mechanism deployed to relate the science of climate change: Gore as both prognosticator and translator.

As Davis Guggenheim relates in the director's commentary associated with the documentary: "Time and time again, things that Al was talking about in the abstract were starting to become real, and that's why we were so motivated to make this film quickly." In part, this role as soothsayer is linked to Gore's insider knowledge; throughout the film, he makes references to "his good friends" distinguished paleoclimatoligist Lonnie Thompson and astrophysicist and author Carl Sagan, or his association with Roger Revelle to signal his intimate relationship to science. He often suggests that through his slideshow, the audience is privy to information never seen outside the rarefied halls of scientific endeavor. In this way, Gore acts as a central node in translation network.[3] Gore is a means of dissemination, through which climate science flows. He translates knowledge of biophysical reality of climate change and works to circulate and simplify the inscription of climate science into graphs, charts, measurements, and scientific studies to us—mobile data that can circumnavigate the globe. In doing so, Gore is a vehicle through which knowledge travels to the layperson, not only allowing for broader community access to climate science but also legitimating himself as an important pivot for its dissemination in the process. It is for this work that Gore won his Nobel Prize in 2007.

Given that one of the key elements of the documentary form is the construction of an authoritative and trusted voice who can, as the saying goes, "speak truth to power," it should come as no surprise that Gore is venerated as both soothsayer and translator. It is Al Gore that compels belief; his authority made through his long-standing connection to the issue of global warming, his political career, and the acceleration of his standing through the film and his attendant climate activism render him a reliable and faithful agent through which climate change can be known. But it is also the science itself, made visible and understandable through Gore's intervention. Thus, the documentary itself acts as another kind of Latour's immutable mobiles, but one with a more fulsome potential to enroll people the world over in its network of truth. Subject and object become one through the personage of Al Gore. Through the use of scientific evidence and the authorization of Gore as an intermediary to translate it, *An Inconvenient Truth* appears a naked representation of the facts. However, as Trinh Minh-Ha (1993), Nichols (1991, 2001), and Renov (1993) have remarked with regard to documentaries more generally, the seemingly transparent process by which knowledge is translated in this filmic form is anything but. The gathering of evidence; its articulation with relation to other kinds of evidence; the selection of particular images, language, and concepts; and the occlusion of others make the production of an argument a creative act, rather than a revelation of preexisting truth. Nevertheless, Gore's journey through the scientific basis of climate change makes it appear to be a clear exposition of fact.

And so, for *An Inconvenient Truth*, the most potent weapon in the war on climate change is scientific knowledge, as translated by Gore. What is interesting, then, are the moments of discontinuity, the cracks in the film's narrative of scientific truth. Occasionally the lines between fact and fiction are blurred. For example, the documentary utilizes a sequence from the major motion picture *The Day after Tomorrow* to illustrate the ice shelves in Antarctica. To be sure, it is a dramatic image, but one that does not hold to Gore's exposition of scientific fact. This is rather a small quibble, but one for which there has been much hue and cry by "skeptics." For them, it seems to illustrate the wholly fictive endeavor of *An Inconvenient Truth*. For me, the inclusion of this scene speaks rather to the quest to make the film a visceral and compelling representation of climate science, one that utilizes visual tropes already in circulation. Indeed, both *The Day after Tomorrow* and *An Inconvenient Truth* rely on the reports of the IPCC, making the boundaries between the two films more porous than one might assume.

More interesting than the inclusion of this image is Gore's explication of the nature of scientific discovery itself. He alerts the viewer to the notion that science is a series of ruptures in thought, in particular through his discussion of Pangaea at the beginning of the film. The reader will recall that he makes much out of his discussion of his grade school classmate who, upon looking at a map of the world, queried whether Africa and South America once fit together. In Gore's anecdote, his teacher ridiculed his classmate because the geological sciences had not discovered this particular fact. Gore uses this example to demonstrate a lack of scientific imagination; he asserts that the teacher was "actually reflecting the scientific establishment of that time" (*Inconvenient Truth* 2006). Gore argues that the assumption that continents could not possibly move constrained the universe of thinkable thoughts in scientific discourse. Of course, Gore goes on to link this anecdote to another assumption: that the earth is so vast humans cannot possibly change its climate. But what's interesting for me in this anecdote is not so much that it shows the assumptions embedded in a climate change skeptics' position (which I will grant it clearly does) but rather illuminates the contingency of scientific investigation. Since Thomas Kuhn (1996) through to Science and Technology Studies scholars today, we know that the scientific discovery is not the simple and progressive building of fact upon fact until the truth is realized. Gore illustrates this well through his Pangaea example. However, while signaling to the viewer the social nature of the scientific process at the beginning of the film, he takes the truth of science for granted for the rest of the documentary. Of course, it would not make a compelling argument to suggest that there is uncertainty embedded in the scientific project on climate change, or that ruptures can and do occur in the scientific thought on this subject. So while offering a hint of the complexity of scientific investigation (which by no means renders it invalid), Gore retreats to the rather more predicable articulation that science can only speak the truth of the biophysical world.

And yet, Gore's scientific assertions are somewhat complicated through his reliance on religious and moral language throughout the film, which I discuss more fully below. Now, this is not to say that science and religion cannot find union; indeed, to use the words of Bill Nichols, both are "discourses of sobriety . . . [that] are vehicles of domination and conscience, power and knowledge, desire and will" (1991, 3–4). But generally, in the realm of climate change, one does not find the two discourses invoked together.[4] However, Gore continuously puts these two systems of thought in conversation with one another, linking both scientific evidence and moral purpose in common cause. This is a discontinuity that makes the film a more compelling document for investigation.

Heroism and Eco-Gore

Scene: This scene opens with Gore looking out of an airplane window contemplatively. He narrates: "Making mistakes in generations and centuries past would have consequences that we could overcome. We don't have that luxury anymore. We didn't ask for it, but here it is." The scene suddenly shifts to coverage of the 2000 presidential election, where Gore and Bush Jr. vied for office. The music swells and there are images spliced together to tell the story of this contentious election: newscasts, press conferences, the details of the recount, officials examining the infamous hanging chads, and finally Al Gore conceding to the decision of the U.S. Supreme Court to stop the recount. George Bush is sworn in as president, in an election many argue was fraudulent. Gore states, "Well, that was a hard

blow, but [*long pause*] what'dya do? You make the best of it. It brought
into clear focus the mission that I had been pursuing all these years. And I
started giving the slideshow again." And so, as Gore loses his bid for the
presidency, one that ostensibly he had prepared for his entire life, he finds
his true purpose: to tell the world the perils of climate change.

The story of Gore's loss in the 2000 election provides the lynchpin in a
broader narrative scaffolding that works to lionize the former vice president, to transform him from a failed politician into a mythic hero. Indeed,
the perceived injustice surrounding the adjudication of the election results
lends him a kind of moral authority he would otherwise lack. In what
might have destroyed lesser men, for Gore the 2000 election allows him to
shed the shackles of a political career that could only serve to hamstring
his true mission. In doing so, he gains fame (and a good deal of fortune)
but as Davis Guggenheim asserts in the director's commentary, he doesn't
crave celebrity, but rather enlightenment:

> [W]hat happens to a man who invests so much in a political path?
> What is he going to do after that loss? To me, this is the most
> heroic thing about Al. This is the most heroic step in his journey.
> You come so close to winning the presidency and you don't get it.
> So what do you do with your life? And a lot of former politicians
> cash in; they sign a big deal. A lot of politicians would sort of enjoy
> the spoils of being famous and the glamour of a former political
> life. You meet Al today and you say look at the choice he's made.
> Here he is going up on stage to another audience. He doesn't get
> paid; he doesn't go for the glamour of it. He is out there trying to
> tell a story, he's trying to convince as many people as he can that
> this is real. To me, that's why all of us are so inspired by him is
> because of that choice, that heroic choice, that he had made.
> Joseph Campbell, the famous author, the man who George Lucas
> based his *Star Wars* movies over, talks about what a hero is, and
> I'm going to retell it terribly, but the idea is that you have a person
> who overcomes great obstacles and achieves things. If you think
> about Al Gore out there in the world, confronting great obstacles,
> trying to tell the story, taking it to the congress in the '70s, trying
> to get in the platform of the Democratic party, trying to tell the
> story and for many years no one listening. And then losing this big
> election and going off and trying to do this thing and trying to

achieve this thing, trying to break through this barrier and get people
to understand this issue. To me, that's heroic. To me, that's the
path and the journey of a hero. And I think that's why people are
rallying around him. People are saying that he's been rediscovered.
I think the truth is he's always been this way and we're suddenly
waking up to who he really is. (*Inconvenient Truth* 2006)

This likening of Al Gore to a mythic hero has not been lost on scholars.
Indeed, Rosteck and Frentz (2009) have used Campbell's notion of "the
monomyth" to think about the construction of Al Gore in *An Inconvenient
Truth.* They suggest that the presentation of Gore's life in the film works
through Campbell's heroic narrative framework of "departure, initiation,
and return," whereby a common man (almost always in the masculine) is
transformed into someone extraordinary (Rosteck and Frentz 2009, 4). In
the case of Al Gore, this conversion is related to the trials and tribulations
of his life. Alongside his failed bid for the presidency in 2000, we can place
Gore's early exposure to the science of climate change at Harvard, the near
death of his son in a car accident, his unsuccessful attempts to raise awareness around global warming as a senator, and his work as Clinton's vice
president; all of these personal and professional moments marked Gore,
both scarred and enlivened him, so that he comes to a privileged position
of knowledge and truth. This, as Tim Luke has suggested, places Al Gore
at the forefront of the "climate-conscious vanguard" (2008, 1816)—those
individuals, agencies, scientists, and celebrities authorized to educate the
rest of us about the perils of climate change.

And so, *An Inconvenient Truth* is the story of Al Gore as much as it is the
story of climate change, or rather, that the two narratives cannot be separated. Indeed, Rosteck and Frentz point out that Gore *is* the movie in a
departure from the traditional expository documentary form:

For, unlike many conventional documentaries, there are no
interviews; no scientists are directly consulted; no variety of voices
offers insight, comment, or context as we might expect from the
prototypical science documentary. Instead, the film and photo
footage invites us to certify the credentials of Al Gore himself as
we come to understand how he has transformed himself into
someone who is qualified to speak on the basis of his travels and
his conversations—in short, Gore is constructed as "expert" and

is thus able to relay what he has learned from "being there" to the rest of us. (2009, 9)

We witness Al Gore giving the slideshow, strategizing with aides, traveling around the globe, engaging with audience members; he has become both source and interlocutor. Director Davis Guggenheim gives us the sense that Gore is otherworldly, so far above the banalities of everyday life that he exists only to educate. In the commentary associated with the movie, Guggenheim describes Gore as a force of nature, "moving faster than everyone else." He lives only to further his cause, working at what would be for some an exhausting pace. Not so for Gore: "4 a.m. most other people are asleep, Al is still working—whenever he has a moment, he is working—heroic and powerful, him alone working at his computer trying to convince his next audience why this is important" (*Inconvenient Truth* 2006). Toward the end of the film, accompanied by the visual of Gore backstage, preparing to present his slideshow, Gore remarks: "There's nothing that unusual about what I am doing. What is unusual is that I had the privilege to be shown it as a young man. It's almost as if the window was opened through which the future was very clearly visible. 'See that,' he said, 'See that? That's the future in which you are going to live your life.'" And so, the documentary makes Gore into an eco-hero, a mythic figure whose larger than life biography and single-minded commitment render him legitimate, authoritative, and ultimately, the most trusted brand in climate discourse.

Affect: Nostalgia, Fear, and Apocalypse Now

Scene: We are at the Gore family farm, a bucolic image replete with cows, greenery, and dappled sunlight filtering into the barn. The scene is slightly out of focus and saturated with a yellowish hue; the camera used is an 8mm, evoking both simplicity and nostalgia. Gore's voice-over starts, recalling the days when he worked tobacco on the family farm. Standing in the tobacco farm, he relates that although the surgeon general announced in 1964 that there was a link between cigarettes and lung cancer, the Gore family continued to grow tobacco. The narrative then shifts abruptly to his sister Nancy, set among images of her and her younger brother. "Nancy was almost ten years older than me and there were only the two of us. She was my protector and my friend at the same time. She started smoking when she was a teenager, and never stopped. She died of lung cancer. That's one of the ways you don't want to die. The idea that we had been part of the economic pattern that produced the cigarettes that produced the cancer was so . . . it was so painful on *so many* levels. My father, he had grown tobacco all his life. He stopped. Whatever explanation had seemed to make sense in the past just didn't cut it any more. He stopped." We return again to the tobacco farm. Gore tells us, "Its just human nature to take time to connect the dots. *I* know that. But I also know that there can be a day of reckoning when you wish you had connected the dots more quickly."

This retelling of the story of his family connection to big tobacco leaves some details out. For example, it erases the fact that the Gore family wasn't immediately struck by the consequences of this "economic pattern" and, indeed, continued to grow tobacco for a further seven years after Nancy's death. It also omits that Gore's various political campaigns were funded by tobacco money during that same time period (Turque 2000, 161, 309). Gore has further been accused of using his sister's death as a sound bite, an exploitative political strategy, both in his 1992 book *Earth in the Balance* and a speech in the 1996 national Democratic convention, where he recounted the details of his sister's last days to argue for stronger regulation around the marketing of cigarettes to minors (Turque 2000, 308–10). That Gore and his family were affected by Nancy's death goes probably without question. But the point is that this is also a political story, one that Gore deploys in the public sphere to narrate something about himself and his trajectory on the path to enlightenment. It serves as a heuristic to explain one aspect of Gore's mythic journey.

But the story has another purpose in the context of *An Inconvenient Truth*. It works to evoke feeling or to generate affect. This is a broader theme throughout the documentary. As Blair (2008) has argued, feeling is central to the argument of the movie, particularly because climate change lacks both the visceral and immediate character of other environmental issues (at least for some): "Since most of us cannot feel climate change, *An Inconvenient Truth* invites us to feel with and through Gore, and then to transfer this feeling into a reaction directed at the environment" (320). The story of Nancy's life and death is but one element in this larger approach. Gore and Guggenheim rely on nostalgia and fear to induce action on their climate agenda.

Nostalgia is a particularly potent way telling stories about the world. Filled with yearning, lament, and a halcyon sense of a once better time, nostalgia permeates many readings of the past. And nostalgia is no stranger to environmentalist scripts. In Disney's Animal Kingdom, we saw its operation through the imbrication of race and nature, imagining the Global South as the repository of nature unencumbered by cultural trace. In Yellowstone and Grand Teton, the entire ecotour was predicated on this romantic mourning for a nature now lost in modern urban life. We find this sense of loss again in *An Inconvenient Truth*. Robin Murray and Joseph Heumann (2007) argue that this emotion operates on two registers in the film: first through the recounting of personal stories from Gore's life, and second through the collective memory of once pristine earth. These authors suggest that the personal memories that the film tells work not only to generate a feeling of connection to Gore, but also to the earth he seeks to save. The stories Gore relates are always linked to an environmental mission, and so, as Murray and Heumann argue, they are reinforced by the larger narrative of the slide show, working in tandem to produce a coherent narrative of environmental decline. Davis Guggenheim elaborates on this point in the director's commentary, suggesting that the Gore family's experience with tobacco is analogous to our experience with global warming. Gore and his family, Guggenheim contends, are made human through the retelling of his sister's cancer; they were connected to a system that was both profitable and destructive, ignoring the surgeon general's warnings. The same is true for those of us who chose to ignore the warnings of impending climate peril. And so, because we are involved in creating the problem (as was the Gore farm) we are also bound to its solution. Gore's story, one of struggle but also acclaim and renown, teaches the audience lessons about adversity, effort, and redemption. The personal memory relates to the collective responsibility. Thus, emotion and reason (as Murray and Heumann term it), or affect and science, come together in *An Inconvenient Truth* through a nostalgic sense of loss, mourning, and desire.

One vignette in particular captures the sense of loss and sadness but also provides the scientific evidence that Gore seeks to associate with climate change. Here I refer to the computer-generated polar bear looking for ice (see Figures below). Gore explains that the Arctic is heating up faster than any other part of the world, and this has impacts on animals like polar bears that rely on the ice for survival. He recounts that bear drownings have been reported as the Arctic sea heats up. In the animation, the bear—panting with what one assumes is exhaustion—searches for the fast-disappearing sea ice that Gore describes. It finds a small piece, which cracks under its weight. It tries again to the same effect. The camera pulls back

An animation from the film An Inconvenient Truth *depicting a polar bear searching for sea ice that is fast disappearing due to climate change.*

and the vast expanse of the sea is free of any ice. The bear sets off into the horizon and, we are to imagine, to its death.

This segment is heart wrenching. It personalizes the story of climate change, but rather than focusing on people the world over who are already suffering the effects of global warming, the scene relies on the connection the audience will feel to charismatic megafauna like the polar bear. However, it does not depict a real polar bear, which might appear to be both

too graphic for audiences (more like a PETA documentary) and too difficult to achieve cinematically. It also might be impossible to capture, given
the controversy over whether bears are actually drowning. Either way, the
animation does its work. The images generate strong emotion—a sense of
the struggle and nobility of the animal, despair at its fate, and regret that
its life is cut short through human action, potentially our very own actions.
As Guggenheim contends, "watching the movie with an audience, people
are devastated; you can hear the gasps and sighs" (Inconvenient Truth 2006).
This scene evokes mourning and loss, not only for the animal, but also for
its metonymic representation of all wild nature.

Murray and Heumann (2007) contend that cultural critics cannot dis-
miss nostalgia out of hand; this way of apprehending the world, they suggest, isn't "inherently and inescapably retrograde." Instead, they argue that
it can be a vehicle through which people can connect to the climate crisis,
using their memories as the basis to act for change. But I would put forward that what is deployed in An Inconvenient Truth is more than simple
nostalgia (if ever there was such a thing). Rather, as discussed in previous
chapters, the kind of nostalgia proffered in the film can only serve to reinscribe a separation between nature and culture, between past and present,
between us and them. It is also a particular kind of longing: the loss of a
pastoral or recreational nature.

However, nostalgia isn't the only feeling generated in An Inconvenient
Truth. The visual images in this documentary often rely on a different affective register: raw panic. The scene of the polar bear described above is coupled with another discussion of sea levels—equally as striking and devas-
tating, but less about sadness and more about abject fear. As the music swells,
Gore shows animated images of what will happen if sea level rises twenty
feet: large portions of Florida, San Francisco Bay, the Netherlands, Bei-
jing, Shanghai, Kolkata, and Bangladesh all underwater. He saves his most
arresting image (at least for the Americans in the audience) for last. In
hushed tones, he shows New York City inundated by sea level rise (see
Figures below). Comparing the impact of global warming to the terrorist
attacks of September 11, Gore shows before and after pictures of the city,
focusing the audience's attention on the World Trade Center Memorial,
which would be under water if climate change continues unabated. The
water creeps insidiously across the aerial view of New York, changing forever the landscape of this quintessential American city. In light of this dystopian future, Gore asks his viewers: "Is it possible we should prepare for

The World Trade Center Memorial, before and after. An animation from the film
An Inconvenient Truth *depicting the impact that sea-level rise would have on*
Manhattan, causing the World Trade Center Memorial to be underwater.

other threats besides terrorists?" In a country whose politics is often characterized by the war on something (cancer, drugs, crime, terror), *An Incon-*
venient Truth seems to suggest another war is required: the war on climate
change. Anything less would be insufficient to the cause.

But Gore's evocation of dread and war doesn't end with sea-level rise and
the portents of doom hard to deny. While Laura Johnson (2009) suggests

that *An Inconvenient Truth* offers a different kind of rhetoric, more muted that that of Rachel Carson or Paul Ehrlich, I, along with a number of other critics, would suggest otherwise. For example, Nordhaus and Shellenberger (2007), authors of the now (in)famous article "The death of environmentalism," suggest that Gore offers his audience a typically green dystopian future, where life is hard and sacrifice is the norm: "As surely as the Bible begins with a fall and ends with apocalypse, humankind's sins against nature will be punished" (Nordhaus and Shellenberger 2007, 106–7). And Gore shows us the results of our transgression. Repeatedly, the film relies on a visual logic of fear to narrate the story of climate change. As Gore tells us, "it's a nature walk through the Book of Revelations." The disappearance of the Larsen ice shelf in Antarctica, receding glaciers, the possibility that ocean currents will shift, the predictions of mass extinction, the forecasts of increasingly violent weather events, with particular focus on the aftermath of Hurricane Katrina: each of these scenes contribute to the building sense of anxiety that infuses the film. This anxiety works not only on the mind, but generates sensory response: grimacing, a feeling of panic in the pit of your stomach, nervous energy, the passage of tension through the body. The fear is palpable and crests just at the end of the movie, when Gore offers potential solutions to the coming crisis. So, part of the persuasive capacity of *An Inconvenient Truth* is that it weaves its narratives through an elicitation of affect. Each scenario engenders complementary feelings of nostalgia and alarm, a lament for an idealized communion with nature that never was and anxiety at its loss.

The Active Citizen: "Are we capable of doing great things?"

Scene: Gore stands before his audience. He states, "Final misconception: if we accept this problem is real, maybe it's just too big to do anything about. And you know there are a lot of people who go straight from denial to despair, without pausing on the intermediate step of actually doing something about the problem. And that's what I'd like to finish with, the fact that we already know everything we need to know to effectively address this problem. We gotta do a lot of things and not just one." Gore then goes through a range of things we can do to reduce our carbon footprints, both individually and on a broader scale: use more energy efficient appliances, improve vehicle efficiency, utilize renewables, and implement carbon capture and sequestration technology for coal. He shows how the combination of all of these actions will take the United States below 1970 levels for carbon emissions. He goes on to use his now patented phrase:

> We have everything we need, save perhaps political will. But you
> know what? In America, political will is a renewable resource
> [*clapping from audience*]. We have the ability to do this! Each one
> of us is a cause of global warming, but each of us can make choices
> to change that [*images of gridlock and suburbs*] with the things that
> we buy, the electricity we use, the cars we buy [*images of high-tech
> public transit and hybrid vehicles, as well as wind turbines*]. We can
> make choices to bring our individual carbon emissions to zero.
> The solutions ARE IN OUR HANDS. We just have to have the
> determination to make them happen. Are we going to be left
> behind while the rest of the world moves forward?

Gore then goes on to describe the ways that the United States and the world have risen to particular challenges: ending slavery, supporting Nelson Mandela and the anti-apartheid movement, conquering smallpox and polio, bringing down communism, working globally to address the ozone hole. He returns once again to the image of "earthrise" from the Apollo 8 mission. He says, "And that is what is at stake: our *ability* to live on the planet earth, to have a future as a civilization. I believe this is a moral issue. This is your time to seize this issue. This is our time to rise to secure our future."

This extended scene (which comprises two chapters of the DVD) illustrates two interrelated arguments of the documentary. The first is the significance of the active individual in combating climate change; when Gore finally turns to solutions, the green citizen is center stage. This person, as described in previous chapters, is one who has the capacity *to do, to be, to become.* In this final section of the film, Gore issues a clarion call to engage in the small acts of change that can lead to an avalanche. This theme is also demonstrated in the closing credits of the film, where, if members of the audience chose to stay, they learned about potential solutions they could incorporate into their lives. Streaking across the screen are a list of such efforts: purchasing energy efficient appliances and light bulbs; lowering your thermostat; weatherizing your house; recycling; buying a hybrid car; walking, riding a bike, or using public transportation when possible; switching to renewable energy sources and convincing your power company to do the same; voting for politicians who support climate legislation, as well as lobbying congress and potentially running for a seat; planting trees; writing letters to the editor or calling radio shows; praying for change, if you believe in prayer; learning about climate change, and acting on this knowledge; and, of course, encouraging others to see *An Inconvenient Truth.* Gore places a special emphasis on youth in this final section of the film. In the credits, we are told, "tell your parents not to ruin the world that you will live in," and "if you are a parent, join with your children to save the world they will live in."

All of these recommendations offer a particular kind of subjectivity that, as in the previous cases elaborated in this book, largely hinges on a form of green consumption. The prescriptions Gore offers work to individualize and totalize, making the subject feel valorized for her or his small acts for the environment while simultaneously connecting to a broader public comprising enlightened individuals, who, like Gore, see the potential environmental catastrophe that now looms. Through changing light bulbs to compact fluorescents or planting trees, the individual Gore targets in *An Inconvenient Truth* can feel secure that at least they are doing something to forestall the climate crisis. Even if we all know that driving a Toyota Prius won't save the world, we can feel good about doing it just the same. And so, what Gore's blueprint allows for is the potential for incremental rather than revolutionary change to heal the globe—actions and movements that are relatively easy to incorporate into privileged lives. Of course, incremental changes can add up; according to Energy Star, an energy efficiency program of the U.S. Department of Energy and the Environmental Protection Agency, "If every American home replaced just one light with a light that's earned the ENERGY STAR, we would save enough energy to light more than 3 million homes for a year, about $600 million in annual energy costs, and prevent 9 billion pounds of greenhouse gas emissions per year, equivalent to the emissions of about 800,000 cars" (Energy Star n.d.). No small feat. And Gore is right to give people things that they can do immediately, in day to day life, to combat the despair and panic that might accompany the acceptance of climate change as a reality. However, there are limitations implicit in this line of argument. Gore's exhortations suggest that we already have the tools to solve the climate change crisis, and they involve a reliance on technological fixes. We just need to implement them.

Even when *An Inconvenient Truth* addresses larger scale changes like reducing U.S. dependence on foreign oil (rather than reducing oil consumption), increasing electricity and passenger vehicle efficiency, switching to renewable energy, and initiating programs for carbon capture and sequestration (a technology that to date has not proven to be successful), they all remain locked within the notion that the excesses of civilization can be redeemed through the further application of particular brands of science, geo-engineering, and cutting-edge technology. Some say this keeps us locked within the very frameworks that have produced the climate crisis (see, for example, Sachs 1999). However, more importantly, it limits the field of discourse and imagination of possibilities for difference. By remaining within the realm of individual action or technological fix, the lessons offered in *An Inconvenient Truth* erase other questions about the origins of and solutions to climate change that might incorporate questions of equity, justice, and power. Instead, we can approach climate change through a capitalist framework of consumer choice, corporate innovation, and limited government regulation. Really not so inconvenient after all.

But at the same time as the narration of this story that focuses on the individual and her acts of environmental restoration, there is also another about broader communities. And this story, like the others told in this book, is an American one. *An Inconvenient Truth* invites individual audience members to feel not only guilt for inducing anthropogenic climate change, but also empowerment in that there are simple tools to rectify what seems to be the apocalypse in our midst. But, amid both the nostalgia and fear, there is the sense that American ingenuity will find a way (as, purportedly, it always has). This generates another affective dimension: nationalist pride, which is complicated to some degree by the inverse notion of national shame. The shame comes in the form of the abandonment of America's once golden history and promise as a leader among nations. Gore tells the audience that the United States stands with Australia as the only "advanced" nations not to have ratified the Kyoto Protocol. There has been a lack of leadership, an erosion of democracy, and a deficit in political will to solve the crisis, which, at least in environmental affairs, has caused the United States to abdicate its role as a global leader. But embedded within this disgrace is the possibility of redemption. As explored above, Gore recounts the many times that the United States has risen to conquer seemingly insurmountable challenges, from the eradication of slavery to the regulation of ozone depletion. The pride generated through this recitation gives a moral purpose to the United States that Gore feels has been lacking in recent years. And we now stand at another moral crossroads in Gore's view, one that will allow the United States to rise again. As Gore contends in a *Vanity Fair* article published shortly after the release of the film, "America is beginning to awaken. And now we will save our *planet*" (Gore 2006, emphasis in original). In the same article, he elaborates on his point, emphasizing the moral imperative that runs through the entire film:

> This crisis is bringing us an opportunity to experience what few generations in history ever have the privilege of knowing: a generational mission; the exhilaration of a compelling moral purpose; a shared and unifying cause; the thrill of being forced by circumstances to put aside the pettiness and conflict that so often stifle the restless human need for transcendence; the opportunity to *rise*. . . . When we rise, we will experience an epiphany as we discover that this crisis is not really about politics at all. It's a moral and spiritual challenge. (ibid., emphasis in original)

What Gore offers in the movie and its attendant publicity, then, is the opportunity for the United States to reclaim its moral authority on the world stage, to once again to be the harbinger for democratic principles and positive leadership. With this purpose, America will fulfill its destiny. So, what underpins Gore's vision of both the individual subjectivity and imagined community of the nation is the embrace or a moral purpose through which to govern the conduct of each and all. The narrative begins with transgression but ultimately ends with redemption, if the proper ethical choices, as outlined by Gore, are made. And so, Gore offers us mechanisms of green governmentality—a series of techniques, practices, discourses, and subject positions. But unlike those seen in the museum, the theme park, or the ecotour, it works through ethico-political strategies.

The Discursive Setting: 2005–2010

An Inconvenient Truth has received many accolades and much praise as an important vehicle in the war against climate change. However, the impact of a media artifact like *An Inconvenient Truth* cannot be judged only in terms of praise, but also by the degree of opposition raised against it. It is, of course, not the case that Al Gore's documentary simply washed over all Americans, convincing each and all of the urgency of the climate crisis. *An*

Inconvenient Truth did not emerge within a vacuum. Rather, it premiered in a highly partisan atmosphere, where Democrats were far more likely to believe in anthropogenic global warming than were Republicans (see Pew Research Center 2006). Gore himself acts as a lightening rod in these factional debates; he has long roused the ire of those on the political right. For example, the book *Environmental Gore* (Baden 1994) set out to analyze and dismantle Gore's arguments in *Earth in the Balance* (1992), suggesting that Gore was both too apocalyptic and unschooled in the economic realities of our time. This is long before the release of *An Inconvenient Truth.*

However, these allegations haven't gone away with the release of the film. If anything, the barbs have been sharpened. The critiques of both the film and Al Gore generally fall into two camps. The first charges the film with scientific inaccuracies. For example, academics in a recent special issue of *GeoJournal* suggested that the problem with *An Inconvenient Truth* was that, in parts, it exaggerated the science, relying on an emotional rather than fact-based appeal (see, in particular, Legates 2007, Nielsen-Gammon 2007, and Spencer 2007). In a less rigorous and less even-handed forum, the aptly named junkscience.com examines the "misconceptions" of the film, arguing that our climate is always changing and the greenhouse effect isn't necessarily a bad thing. One can find innumerable appraisals such as these on the Web. Questioning Gore's science, and the science of global warming more generally, is not a surprising tactic. Indeed, this approach, along with the suppression and editing of scientific results, was utilized quite handily by the Bush administration as a mechanism to deflect the import of the climate crisis. As Gore himself elaborates, this kind of critique often follows the release of scientific evidence that is bad for corporate interests; he highlights the case of the tobacco industry denial of the addictive nature of cigarettes. No matter the actor—whether legitimate academics, popular debunking Web sites, or government officials—what makes these critiques consequential (if sometimes laughable) is that they offer an alternate reading that seeks to unmask a distorted truth. The fact that a film like *An Inconvenient Truth* warrants such efforts is noteworthy in and of itself.

The second line of critique functions again on truth and attacks Gore as the vehicle through which all climate change propaganda flows, suggesting that there is some grand conspiracy at work to convince the population that climate change is real, that it is produced by humans, and that scientists all agree about it. For example, *The Great Global Warming Swindle*, a British documentary that played on Channel 4 in March of 2007, is a polemical assault on Al Gore's assertion of a scientific consensus around climate change. The film suggests that anthropogenic climate change has become the new orthodoxy, paramount to a religion that cannot be questioned. The documentary argues that, in reality however, climate change occurs naturally. Drawing on interviews with former members of the IPCC, academics, and economists, the film asserts that climate change discourse is political ideology rather than scientific fact, an industry unto itself. Although *The Great Global Warming Swindle* has been dismantled as misinformation (see, for example, Jones et al. 2007; Monbiot 2007; see also realclimate .org), it attracted 2.5 million viewers, and among skeptics it has become something of a cult classic. Similarly, on Fox News in May 2006, Sterling Burnett, who was then a senior fellow at the conservative National Center for Policy Analysis stated, "You don't go see Joseph Goebbels' films to see the truth about Nazi Germany. You don't want to go see Al Gore's film to see the truth about global warming" (Media Matters 2006). Interestingly, the National Center for Policy Analysis had received $300,000 from Exxon-Mobil since 1998 (Lequm 2006). But Burnett wasn't alone in his assessment. Glenn Beck, then with CNN Headline News, stated in June 2006: "[W]hen you take a little bit of truth and then you mix it with untruth, or your theory, that's where you get people to believe. . . . It's like Hitler. Hitler said a little bit of truth, and then he mixed in 'and it's the Jews' fault.' That's where things get a little troublesome, and that's exactly what's happening [in *An Inconvenient Truth*]" (Media Matters 2006). And so, what this second vein shows more clearly than the first is the anxiety around a (perceived) consensus developing on this issue. *An Inconvenient Truth* provoked so much ire, consternation, and angst in part because it seemed to be *working*—to be changing hearts and minds along the lines explored above. How persuasive is it?

The Public Response

Public response to a movie is a tricky thing to gauge. One can look at box office numbers, of course, in which case *An Inconvenient Truth* ranks quite well compared to other documentaries. However, actual impact is difficult to assess from such numbers, for they say nothing about how people felt about the movie and what, if anything, they did with the knowledge they gained. The same is true for awards received. To get at the potential long-term effects of the film, I have chosen to look at three potential indicators: a survey of public perceptions, the initiative called the Climate Project, and the ways that the film has been adopted in educational settings.

Surveying Response

The Nielsen Company and the Environmental Change Institute at Oxford University conducted an online survey in April 2007 to determine who and what were significant actors in communicating information about climate change. The survey covered 47 countries and included 26,486 respondents (Nielsen Co. and Environmental Change Institute 2007a). The survey found a number of things that are noteworthy in assessing the impact of both Al Gore and *An Inconvenient Truth*. Al Gore ranked first globally, with 18 percent of respondents indicating he was an influential campaigner for climate change. Broken down a little further, almost 50 percent of respondents in Belgium and the Netherlands ranked Gore first, along with 40 percent in both Norway and Sweden, 33 percent in Switzerland, 32 percent in Denmark, and 30 percent in the United States (ibid.). Indeed, according to the survey write-up, "A 'dream ticket' for climate ambassadors would include Al Gore and [former United Nations Secretary-General] Kofi Annan, who polled as first or second choices in the most countries, together covering 34 of the 47 countries in Nielsen's Internet survey" (Nielsen Co. and Environmental Change Institute 2007b).

An Inconvenient Truth also scored high with the respondents of the survey. Twelve percent of respondents globally had seen the movie in April 2007 (Nielsen Co. and Environmental Change Institute 2007a). This number was even higher in the United States, at 17 percent (ibid.). The joint survey asked those who had seen the movie a range of questions. The first was whether *An Inconvenient Truth* made them more aware of the issue of climate change. The numbers were then broken down by region: in the Asia-Pacific region, 94 percent agreed; North America—87 percent; Latin America—86 percent; Western Europe—85 percent; Eastern Europe, the Middle East, and Africa—81 percent. The global average for this question was 89 percent (ibid.). The survey then asked respondents if they had changed their habits as a result of seeing the film. The results for this question were slightly lower, though still quite high, with a global average of almost three of out four, or 74 percent, indicating that they had in fact changed some aspects of their lives (ibid.). As Max Boykoff suggests, "*An Inconvenient Truth* has pushed Al Gore and the message of concern for climate change up the public agenda. This has been combined with UN scientific reports and the Stern Review as well as increased media coverage over the last months to shift the focus for many people from whether there is a problem to what to do about it" (ibid.).

This survey seems to indicate a few of important things. First, it suggests the media- and celebrity-driven nature of political and social causes, something that Max Boykoff and others have taken up in different forms (Boykoff 2007; Boykoff and Goodman 2009). For example, along with Al Gore, other ambassadors of climate politics named in the survey were Oprah Winfrey, Angelina Jolie, Bono, Richard Branson, and David Beckham, to name a few. As Boykoff and Goodman contend, these celebrities are "a heterogeneous and important community of non-nation-state actors that have increasingly acted to influence various facets of the science–policy–public interface over the last two decades" (2009, 399). Celebrity, and the attendant spectacle that it engenders, now moves political agendas and environmental causes in interesting, and often, unpredictable ways. Second, it is clear from this survey that *An Inconvenient Truth* has reached large numbers of people around the world. But it appears that people have not just seen the film, but have also been put into motion because of it. For a significant majority of viewers, the response to the film has been to act, to change their lives in particular ways that correspond to what they learned in the movie. I suspect that this is because the film is so successful in its narrative devices highlighted earlier in this chapter: the veneration of Al Gore as truth-teller, the invocation of both science and affect to tell the story of climate change, and the potential subjectivity that film offers its viewers. As this survey demonstrates, *An Inconvenient Truth* works as a discursive object that circulates, facilitates, and promotes a particularly compelling way of interacting with the climate crisis.

The Climate Project

Another interesting way of assessing the impact of the film and of Al Gore as a public vehicle for climate politics is by looking at the Climate Project, an initiative of the Alliance for Climate Protection, the umbrella organization that Al Gore founded to work on various aspects of the climate crisis. The Climate Project was established in 2006 as a means to educate the public about climate change and encourage them to act on this knowledge. It fulfills this aim by training presenters to give an updated version of the now famous slideshow from *An Inconvenient Truth*. Since it's inception, it has trained 3,125 volunteers and has branches in the United States, Canada, Mexico, Australia, India, Spain, the UK, Indonesia, and as of June 2010, China. According to the Web site, these presenters—trained personally by Gore—have given over 70,000 presentations to 7.3 million people worldwide (http://theclimateproject.org).

This grassroots effort at climate education relies again on both Gore's celebrity and his currency as a truth-teller. The opportunity to be part of *An Inconvenient Truth,* to be trained by Al Gore to give the very slideshow that has sparked such concern, is an appealing and complicated one. Gore appears in all his equanimity as the mythic hero who passes on the torch to those who would take up his cause. Indeed, the Climate Project seems to be a factory for Al Gores in training, with each presenter working as a node in the circulation of Gore's message, acting as his agent in thousands of venues one person simply cannot reach. The Climate Project allows Gore's slideshow to continue to circulate and to reach as many people as possible with its message of fear and hope. As such, the slideshow becomes a truth-telling artifact in it's own right, capable of moving from place to place without the need of a specific interlocutor. In a sense, then, we are all (or have the potential to be) Al Gore.

Building Green Kids: An Inconvenient Truth *in Curriculum*

Another venue for the dissemination of Al Gore's message is through high school science and civics curricula. While it is difficult to assess how many schools may have integrated the film into their classes, the National Wildlife Federation, in concert with Gore, has put together a curriculum package that works with U.S. national standards for teachers to use in their classrooms. The curriculum is downloadable for free on the Web site for *An Inconvenient Truth* and is meant to be used in conjunction with viewing the film.

The fully developed curriculum has three tiers geared toward grades 9–12, with each building on the last. Each of the sections is complete with objectives, activities, a list of national standards addressed, and tools for self-evaluation. Tier One is named "The Green Mile to School" and offers an analysis of the relative impact of greenhouse gas emissions, both from a personal standpoint and with reference to different countries, states, industries, and cars. Students are split into groups to research and make presentations on a range of scales that interact in the production of climate change: continental, national, state, regional, and corporate (with focus on the automobile industry). After the presentations, students are asked to evaluate the knowledge they have gained as well as their own participation in the generation of greenhouse gases. Tier Two is called "Think Globally, Act Locally." In this section, students look at the emergence of the Kyoto Protocol and its implementation, with specific reference to the fact that the United States has not ratified this piece of international environmental regulation. Then students look at the actors that have proceeded on Kyoto, including cities across the United States. In the culminating activity, students engage in role-play, acting out the various positions of countries involved in the debate, such as the United States, China, the European Union, Russia, India, and Japan. The activity suggests that local leaders like school board members and city council members be invited to watch the role-play. Finally, Tier Three, named "Smaller Steps Mean Smaller Footprints," turns to students' own lives and community problem solving to examine the impact of climate change. This unit involves a discussion of the movie and scientific modeling on topics like the carbon cycle, both of which lead to brainstorming about the causes and effects of climate change in their own communities. The result of this work is that students come up with action plans that will be presented to local leaders.

While this climate change curriculum offers students a wealth of opportunity to learn about the science and politics of climate change, it also does something else: it serves as a vehicle to build moral green citizens. Striking at students early and working with Gore's focus on youth, the curriculum allows for the transmission of Gore's essential message throughout formal education, lending it both legitimacy and breadth. As a disciplinary institution replete with a range of technologies of subjectification, the school offers the perfect incubator for the making of environmental citizens who will take up the cause of climate change. And the inclusion of authorities like PTA members, city councils, or school board trustees offers the opportunity for this knowledge to generate into action, to exceed the boundaries of the school itself. In this way, the curriculum accompanying *An Inconvenient Truth*, like the Climate Project, provides another means of circulation, but one that is simultaneously more compelling while also addressing a captive audience.

In an interesting turn, however, the adoption of *An Inconvenient Truth*'s curriculum has not been seamless. In 2007 a UK court ruled on a complaint that the film should not be shown in British schools. The judge found that, although the film largely held to scientific fact, there were errors in the film deployed within "the context of alarmism and exaggeration" (High Court Judge Michael Burton, qtd. in Baram 2007). For example, the idea that sea level rise of twenty feet was possible in the near future was questioned. So was Gore's elaboration of glacier melt on Mount Kilimanjaro, the recession of Lake Chad, and the heart-wrenching notion of polar bears drowning. These, among other "exaggerations," were named as scientific problems in the film that gave *An Inconvenient Truth* an unnecessarily apocalyptic tenor. As such, the judge ruled that the film could be used in classrooms, but had to be accompanied by a corollary that the film is not a neutral analysis of the issue. Similar challenges were also raised in Washington State, where a parent who supported a creationist view of history complained that "[c]ondoms don't belong in school, and neither does Al Gore" (McClure and Stiffler 2007). A similar compromise was reached: if teachers chose to show the film to their students, it had to be done alongside the presentation of opposing views. And so rather than named truth (inconvenient or otherwise), the film, at least in these districts, is labeled as a vehicle of propaganda. What these cases show is that although *An Inconvenient Truth* has a wide impact and the ability to shape ideas, practices, institutions, and politics, this impact is not uncontested.

Refining the Message: Our Choice

Al Gore did not finish his campaign for awareness about climate change with his wildly successful documentary. He has continued his labors through the Alliance for Climate Protection and its attendant projects like Repower America and the WE campaign. He has also written a new book called *Our Choice: A Plan to Solve the Climate Crisis,* which was published in 2009. The book looks quite similar to the one that accompanied *An Inconvenient Truth,* using striking images, scientific diagrams, heroic people, and simple prose. But its purpose is different. Gore sets out to pick up where the film left off, arguing that we do have the means to solve the climate crisis, we just need to implement the solutions. Thus, *Our Choice* aims to present these possible solutions: "That's why I have written this book, chosen the pictures, and commissioned the illustrations—to gather in one place all of the most effective solutions that are available now and that, together, will solve this crisis. It is meant to inspire readers to take action—not only on an individual basis but as participants in the political processes by which every country, and the world as a whole, makes the choice that now confronts us" (Gore 2009, 15)." Gore articulates that there are three interlocking crises that are caused by the same problem and thus need to be solved together; the security, economic, and climate crises, he asserts are all linked to the addiction to carbon-based energy (21).

As such, he outlines a series of options along with their benefits and drawbacks, such as solar, wind, geothermal, biofuel, and nuclear power, as well as technological fixes like carbon capture and sequestration. In the end, he suggests that we need to address population, focus on energy efficiency, implement technology like the supergrid to allow this to happen, and elicit a change in mindset, where people embrace that these problems are linked and in need of comprehensive solutions. Gore continues his focus on the individual in *Our Choice,* but also suggests that much more is required—namely, "concerted global action" (2009, 18). And yet, Gore doesn't abandon capitalist principles, although climate change is named a market failure. Instead, we see the same kind of nationalist fervor and market-based solutions found in *An Inconvenient Truth:* "The United States of America's stunning success over the last 200 years (emulated by aspiring democracies on every continent) and the dominance of market capitalism in most of the world (especially after its philosophical victory over communism in the late 20th century) both serve as evidence of the unprecedented power and vitality of these two designs based on the assumed primacy of reason in human affairs" (301). But Gore also argues that the market is not enough; what we need to solve the climate crisis is policy, either a carbon tax (which he supported in the Clinton administration), cap and trade legislation, or direct regulation of emissions through the Clean Air Act. These policy interventions, within a framework of free enterprise as constrained at least partially by the state, are what Gore sees as the path toward ending the climate crisis. This approach is also emphasized in his TED talk,[5] where he argues that while it is important to change light bulbs, there needs to be a broader embrace of political and democratic citizenship. However, he continues his veneration of technology and individual action, suggesting that the way to "go far quickly" is through the power of information—using technology to help visualize, model, and monitor the climate crisis, which, Gore argues, will lead to new kinds of knowledge and practices. Along these lines, he advocates for the use of the Google PowerMeter, which allows consumers and companies to monitor their energy usage in real time (recall that Gore is a member of Google's corporate board), the use of telecommuting, congestion charges, and other technology-based mechanisms that would alter lifestyle, raise awareness, and lead to change. However, all of this focus on policy, technology, and concrete solutions makes the conclusion of the book rather discordant. In the end, the book returns more to *Earth in the Balance* than *An Inconvenient Truth,* as Gore embarks on an extended thought experiment or utopian epilogue for his readers. He writes how he would like to see the future unfold as if the United States has already "awaken[ed] to its responsibilities" and implemented Gore's vision for a green politics and public (399). He then takes a religious turn toward the end of his conclusion, which struck me as so peculiar that it is worth quoting at length:

> With God as our witness, we made mistakes. But then, when hope seemed to fade, we lifted our eyes to Heaven and saw what we had to do. . . . Seen from the vantage point of space, our planet is the Garden of Eden for all humankind, both living and yet to be born. In our time, without realizing it at first, we attained the knowledge and the power to destroy it. For us then and you now—once again as in the ancient scriptures—the issue is whether we have the wisdom and self-restraint to avoid that outcome, and whether we should use them. The choice is awesome and potentially eternal. It is in the hands of the present generation: a decision we cannot escape, and a choice to be mourned or celebrated through all the generations that follow. (404)

What seemed initially so odd made sense within the larger arc of Gore's mission to combat climate change. It speaks to the moral thread that underpins all of Gore's interventions. Indeed, in an interview with Larry King to promote his new book, Gore expresses this ethical dimension quite fervently:

> If those of us alive today just took the benefits of all the work and sacrifices of previous generations and fully exploited them in our lifetime and gave the back of our hands to those who come after us, it would be the most immoral act of any generation that has ever lived. . . . Most importantly of all, this is a moral issue, not a

political issue. The scientific community is saying to everybody in the world alive today we can't continue putting 90 million tons of this global warming pollution into the atmosphere every day without risking an unprecedented catastrophe that could threaten the future of human civilization. (*Larry King Live* 2009)

This focus on morality is where I would like to end this chapter.

Conclusion: Moral Governance

Like the museum, the theme park, and the ecotour, I assert that *An Inconvenient Truth* relies on the impartiality and unquestioned truth of science to warn of impending global apocalypse, reinscribing the preeminence of this way of understanding nonhuman nature. But more than its appeal to the scientific endeavor, I contend that this film also works through narrative—through the affective potential of storytelling. Through science and storytelling then, *An Inconvenient Truth* generates a script of improvement: the technologies necessary to the remaking of the self for the best end of both human and nonhuman nature. Relying on affect and knowledge—heart and mind—Gore builds the perfect green citizen. Thus, Al Gore's film is *productive*, like the other sites examined in this book.

What is unique in Gore's project is that in using some of the same registers as the previous sites, Gore offers the viewer a kind of *moral governance*, telling us how we can be ethical human beings in a time of environmental crisis. Part of the appeal of *An Inconvenient Truth* is that it offers the viewer the right way of living, a toehold on a moral way of encountering nature. The film tells us that we are in the grips of a climate crisis and that the responses of the U.S. government and citizenry have been unethical, profligate, and self-interested. Here Gore stands as the paragon, deploying the affective dimensions of nostalgia, fear, citizenship, and nationalism, as well as science and celebrity, to tell the story of environmental declension and crisis. But in the narration of this crisis, with its images of polar bears drowning, New York City and Kolkata engulfed by the ocean, and scenes of mountains laid bare, denuded of their glaciers, Gore offers the potential for redemption, both for individuals and as part of a national and global citizenry.

What is equally compelling about this brand of morality is that it isn't too difficult to incorporate into our lives. There are relatively simple practices and, as Gore asserts, we already know what we must do. Energy efficiency or taking public transit are among the means to achieve a moral purpose, a sense of anchor in a world increasingly characterized by crisis and anxiety. What Gore's story makes clear is that governmentality can take a very individual form. He directs his crusade at each and every person by arguing that you—the watcher and listener—must take the ethical stance, must become a green citizen. Gore acts as a moral guide who leads the willing subject to an improved self: he provides the vision, he elaborates the conduct, he specifies the exercises, and he explains why and how you can become better. That these means to ethical subjecthood are often rooted in consumer choices should come as no surprise; indeed, this brand of green governmentality is analogous to many of the techniques of power seen throughout the rest of this book.

What makes *An Inconvenient Truth* both interesting and unique is that it embodies a kind of green moral governance in one person—the Goreacle. Unlike the museum, the theme park, and the ecotour, what matters in *An Inconvenient Truth* is not just the biophysical reality of climate change, but also the credibility of its spokesperson. So, the film is compelling because of the equal weight it gives to both science *and* storytelling. Indeed, much of the film works to establish Gore's character and authority.

It amounts to a kind of confession of past sins and blindness from which his new mission provides salvation. Gore and director Davis Guggenheim suggest that his personal and family troubles fostered an "examination of conscience," moments of epiphany through which Gore had his vision cleared so he could see the crisis the rest of us have ignored. Of course, Gore is acting out a ritual of our times that focuses on confession. In popular culture, one can look to Oprah Winfrey for confirmation of this trend, but it is also seen, for example, through the proliferation of reality TV shows like *Intervention,* the craze around blogging and tweeting, or the fact that many sat rapt in early 2010 as Tiger Woods held a press conference to contritely admit his marital misdeeds. But this focus on public redemption goes far beyond popular culture; it permeates diagnoses of mental, physical, and spiritual health, the justice system, and practices of sexuality. As Renshaw argues, "To fail to know and be able to confess/speak your 'inner truth,' in Western culture today is not to experience an existential crisis but an ontological one" (2010, 175). We tell the truth of ourselves to both convince others of our worth and remake ourselves anew. Gore's intervention can be placed within this confessional framework. *An Inconvenient*

Truth offers Gore as the green version of the "confessing animal"; through the confession of his life story—his mistakes, missteps, and early conversion—he is transformed into the redeemed moral subject. However, this is not the kind of confession that induces guilt, shame, or despair. Rather, it is productive. It produces Gore as an archetype for the kind of truth-telling he suggests each of us must employ to confront the climate crisis, rooted in an examination of self and past practice that leads to enlightenment and change.

But Gore's moral truth-telling isn't as simple as all that; while he carries a good deal of ethical weight, he is not an unblemished mythic hero, despite what Davis Guggenheim and others might contend. If we turn to Foucault's lectures at the University of California, Berkeley, in 1983, compiled into a book entitled *Fearless Speech* (2001), we can see that Gore is a more complex hero than he might appear. *Fearless Speech* deals with the conditions of possibility for the emergence of truth-telling, or *parrhesia,* in the Greek world. The lectures address "the importance for the individual and for the society of telling the truth, of knowing the truth, of having people who tell the truth, as well as knowing how to recognize them" (Foucault 2001, 170). Foucault traces the emergence of the concept through a variety of Greek texts, from Seneca to Socrates, suggesting that there are a variety of practices of truth-telling (community, individual, and public) that shape relations between self and others. But what do the ancient Greeks offer us to judge truth-telling today? Foucault offers us a series of "tests" to assess the validity and moral importance of the truth-teller, which can usefully be extended and complicated in their application to Al Gore as modern green truth-teller.

The first test is *frankness:* does the truth-teller "open his[6] heart and mind completely to other people through his discourse" (Foucault 2001, 12), and in doing so offer a means to convince others of his views? On the surface, Gore seems to fit the bill. He lays bare both the scientific arguments around climate change and his relationship to them. He puts forward a serious challenge to the way people behave. And he is sincere in his effort to effect real change on the issue of climate change. What diverges here from Foucault's articulation is the use of both propaganda and image that animate *An Inconvenient Truth;* from the nationalist impulses to the sentimental scenes, it is shot through with rhetoric, something that Foucault suggested was antithetical to the practice of truth-telling. Indeed, in Foucault's articulation, this would render *An Inconvenient Truth* a mechanism for ideology, neither transparent nor employing the plain speech necessary to the truth-teller. However, this is where the Greek articulation of *parrhesia* (and Foucault's elaboration on it) doesn't square with modern America. Gore is forthright in his film, but he necessarily employs the tools of our time to tell his story. For unlike the Greeks, we live in a media-saturated world. Gore's use of such technologies doesn't render his narrative less truthful, but rather gives it more force. *An Inconvenient Truth* is a spectacle, to be sure, but this does not mean that it lacks honesty. In this way, Gore offers a modern version of what it means to speak truth to power.

The second test is *risk:* does the truth-teller face danger (physical, political, or emotional) by telling the truth? Is courage required? Is there a tyrant to face or a career to be lost? Gore's role as a truth-teller by this test is a little more ambiguous. As his biographies claim, he seems to risk little in the face of political peril. His environmental record as vice president seems to bear this assertion out. In fact, it is in the hour of Gore's defeat that he finds his role as a hero, that he has been stripped enough to claim the title of truth-teller. He certainly has not risked his life or livelihood as a champion for climate discourse. Indeed, he continues to amass a personal fortune through investing in his ideas. However, Gore has opened himself up to caricature, controversy, and debate through *An Inconvenient Truth* and his books, articles, and public lectures. This is a form of risk enacted through character assassination, where not only his ideas and political record but also his failed marriage and utility bills are up for grabs. His status as a moral hero, which excites his followers but infuriates his critics, leads to this desire to destroy his character—likening him, for example, to Goebbels—to blunt the moral force he might exercise over American conduct. And so, there is danger here and some degree of courage, though likely not the kind that the Greeks imagined in outlining the risk the truth-teller accepts to offer a new way of living.

A third test is *criticism:* this is the means through which the truth-teller names that which is wrong, either in himself or in others. As Foucault asserts, it "may be the advice that the interlocutor should behave in a certain way, or that he is wrong in what he thinks, or in the way he acts, and so on" (Foucault 2001, 17). Again, Gore's position as a truth-teller with regard to this test is somewhat complex. Gore is quite clearly telling a morality tale, a condemnation of excess. We use too much, produce too much, buy too much, and have too many babies. He criticizes the American political system and a lack of will on the part of the citizenry to do something about the climate crisis. He does provide solutions that, if adopted, would present a dramatic change in the way consumer society operates. But his critique is blunted by both his veneration of American ingenuity and his reliance on the marketplace as the key venue for environmental change. And so, while he offers criticism, it is necessarily limited.

The fourth test is *duty:* "a sense of moral obligation" to speak truth (Foucault 2001, 19). This Gore offers viewers of *An Inconvenient Truth* or readers of his books in spades. Gore is imbued with a moral responsibility to educate and through this awakening passes on his duty to save the planet to each and all. By his account, since his early political career he has felt this obligation, and his commitment to it has only heightened since Gore left politics. This is what Gore *does*. And his vision takes on a spiritual dimension; he acts as both prophet and proselytizer, preaching his vision of a new green world and training others to do the same through the Climate Project. Gore is clearly bound by a moral and ethical desire to change the world.

Finally, the last test is *truth:* does the truth-teller lead a life of conviction, where practice corresponds to words? Is there harmony between what one does and says? Does Gore lead "the true life"? The answer here seems to be a resounding, well, kind of. Gore's personal habits have been a matter of much debate, as one might expect given he offers a right way of living for the rest of us. In many ways, he does back up his talk with action. All the proceeds from the film and his books go towards his various nonprofit organizations, like the Alliance for Climate Protection. He invests in green technologies. He travels the globe to preach his message and offsets this travel. He purchases green energy and has installed solar panels for his twenty-room mansion in Nashville, accelerated perhaps by an exposé that revealed he used $30,000 worth of energy in 2006 (Leonard 2008). And yet, something about a green crusader living in a mansion seems to stick in people's craws. Because, of course, part of being green is not living in sprawling suburbs with swimming pools, or flying all over the world to give talks about climate change. It certainly is not a "small is beautiful" mantra of organic farming or living locally. However, his rather lavish lifestyle, at least by middle-class standards, is not necessarily incongruent with his larger message that might be summed up as live smart, not live cheap. So, Gore does adhere to his own standards of truth, where capitalism and environmentalism find happy union.

In the end, *An Inconvenient Truth* and the persona of Al Gore offer us an ethics of the self, an individualized form of green governmentality. While Gore does not necessarily receive passing grades on all of Foucault's tests for *parrhesia,* he does act as a preeminent truth-teller, reshaping the categories to meet the needs of the modern world. He holds a kind of moral author-ity—confirmed by such vaunted institutions as the Nobel Foundation—which allows him to act as a guide in matters of green governmentality.

Of course, the debate around climate change is not static and the ter-rain of the discussion is continuously evolving. "Climategate" makes this obvious. Al Gore puts forward a truth claim about the climate crisis, but it is not the only one that circulates, and its hold on truthfulness, while clearly strong, has perhaps been made tenuous since the buzz around *An Incon-venient Truth* has died down. The election of President Obama in Novem-ber 2008 seemed to signal a change in climate politics, but the American political scene has remained a distraught public sphere, where the emer-gence of right-wing populism and the mood of conspiracy have heightened. So, while Gore's narrative is coherent, persuasive, and imbued with moral purpose, it ran against the political exigencies of an electorate more con-cerned with an economic crisis than an environmental one. The results, or nonresults, of the Copenhagen Conference of Parties in December 2009 clearly show this shift. Moreover, it appears as though Al Gore's moral force may be waning, at least according to a 2009 Gallup poll that indi-cates that 41 percent of Americans now think that global warming is exag-gerated (Gallup 2009). One can pin this figure on a number of causes, from the recession to a renewal of vitriolic partisanship, from new envi-ronmental issues like the BP oil gush to the demands of a media-saturated culture that focuses on what's happening now rather than ten or even fifty years in the future. But at least part of this shift can be linked to the vari-ous slights on Gore's character that this chapter elaborates. As a moral figure, Gore has taken his knocks, from *The Great Global Warming Swin-dle* to charges that he will profit from the climate crisis. Perhaps it is because of the sustained attack on his character and credibility that his moral author-ity has lessened, and his ability to offer ethical government to the rest of us has paled. Because, of course, the credibility and character of the truth-teller are crucial to his or her ability to critique the rest of us.

What it also suggests is that it is a tricky business to invest so much in one figure who acts as a green moral compass for us all. What if it turns out that he has feet of clay? So, Gore is a truth-teller, but even his version of the truth, rooted as it is in a faith in the American way, technology, and the market, may not be palatable enough to sustain action on this environmental issue. However, I remain firmly committed to the notion that, although there is never one truth, there are better stories to tell. Al Gore's interweaving of science and storytelling is one way to narrate environmental crisis, and a potent one to be sure. But what other stories can we tell, that might give attention to justice for both humans and nonhumans? This is the business of the final chapter.

Being Otherwise

HESE DAYS IT IS A TRICKY BUSINESS to critique efforts at environmental regulation. Given the predictions of climate change, biodiversity loss, and species extinction, it seems impolitic to challenge programs that appear to ameliorate human effects on the world. In many respects, those of us—myself included—who are environmentally minded find ourselves ensconced in somewhat of a George W. Bush binary: "you are either with us, or against us." As a friend often reminds me, Greenpeace, however problematic, is not BP. Thus, a hazard in my project is that one can be placed alongside those who would deny the human impact on the biophysical world, who would suggest, for example, that climate change is a fiction. This is a side of the fence where I would never want to find myself. And this is the dilemma, in part, with a methodological approach that uses a strict Foucauldian view of discourse analysis. As I signaled in the introduction, I do believe that there are better stories to tell, more just ways to narrate the interaction between human and nonhuman nature.

Moreover, given the urgent nature of these problems, what is the purpose in examining other stories of nature/society relations, such as the ones that have been the focus of this book? Asked more succinctly, are these the important stories to tell? I think they are. While the narratives of the American Museum, Disney's Animal Kingdom, the ecotour to American national parks, and *An Inconvenient Truth* may not directly generate climate change, neither are they separate from it. What each of these cases does is limit our ways of seeing nature, affecting how we can understand and act upon issues of environmental change. In this way, the stories told have discursive and material impacts, asking us to see some aspects of environmental action as up for debate (for example, shopping responsibly), while others become unsayable (for example, the corporate role in environmental issues). By limiting the field of possibility to talk about the environmental crisis, these sites become centrally important in defining what environmental problems are, and hence, what action might be taken to prevent them.

And so, my critique of these places and their function as agents of green governmentality should not be read as a detachment from environmental concern; my goal here is not to suggest that if there is no space outside of power, then nothing is worth doing. In fact, I wish to suggest precisely the opposite. It is in the very act of critique that we might find different definitions of what we come to know as nature, different ways of encountering it, and different means to make this encounter more just. Foucault has remarked: "My point is not that everything is bad, but that everything is dangerous, which is not exactly the same as bad. If everything is dangerous, then we always have something to do. So my position leads not to apathy but to hyper- and pessimistic activism" (1984, 343). If we see the sites I have described in this book as agents of power, nodes in a web of different iterations of green governmentality, then, as Foucault proposes, we have a place to go from here. The difficult part is figuring out what this place might look like.

This conclusion seeks to bring the cases explored in this book together, drawing out their resemblances and differences, demonstrating how each serves to generate similar knowledges, practices, and subject–positions that govern relationships between humans and nonhumans. But I also wish to imagine the ways we might conceptualize this relationship differently, to reveal those stories that cannot be told if we remain comfortably rooted in a truth game where the moves have already been defined. Thus, instead of closure I offer an opening—asking questions about the possibility of an "outsideness" to green governmentality—in hope that there are alternate, and more environmentally just, ways of encountering the nonhuman.

Continuities and Discontinuities
Discourses of "Truth"

The most readily apparent connection between the four case studies examined is that they function to narrate the environmental crisis through an existing regime of "truth": scientific discourse. This is, of course, most obvious at the American Museum, whose purpose is to generate and disseminate the knowledge that scientific observation brings forth. The Hall of Biodiversity is a monument to the will of science to name, classify, and organize nature, depicted both spectacularly and poignantly with "The Spectrum of Life," "The Crisis Zone," and the "Dzanga-Sangha Rainforest Diorama." The museum's explanations of, and solutions to, the biodiversity crisis further rely on the weight of science to narrate nature's catastrophe, more specifically through a discussion of the "Sixth Extinction" and ecosystem services. Science suffuses the fabric of this museum exhibition, giving weight to the stories told and making their "truth" potent and irrefutable.

But the museum is not alone in its reliance on science to tell the "truth" of nature. Disney also deploys science as a means of truth-telling at the Animal Kingdom. Recall the various "research stations" in the attractions of Africa and Asia, the fictional Dino Institute and the trip back in time to rescue dinosaur specimens from extinction, and the whole of Conservation Station. Moreover, Disney's Wildlife Conservation Fund, as well as its research through the Animal Programs department, both point to a new role for this cultural heavyweight: Disney has moved from entertainment to edutainment, using science to both lend itself legitimacy and generate knowledge about the natural world.

The import of science was also present on the ecotour to Yellowstone and Grand Teton National Parks. Our guide was a biologist who recounted the characteristics and typical behaviors of the animals we viewed, as well as the landscape features, in great detail. He spoke about fire ecology and ecosystem management. He hammered home, again and again, the importance of a scientific understanding of nature to the production of "good" environmentalists. More than simply watching animals, the ecotourists were meant to delve deeper to understand how animals behave, communicate, interact, and ultimately, how they are threatened by human action. First, knowledge must be gained; then true environmentalism can follow.

And of course, Al Gore's attempts to educate about the climate crisis are suffused with scientific authority, with visual and affective manifestations of this power in predictions, graphs, charts, and personalities. In the film *An Inconvenient Truth* and his books, articles, and speeches, Gore deploys scientific expertise as the cornerstone in his effort to enlighten; through science the "truth" can be apprehended and acted upon.

In their different ways, each of these sites uses science to lend coherence, authority, and weight to the stories they tell. They function through a preexisting regime of "truth" to assert their own stories about the fate of the environment. Through the use of scientific discourse, it can appear as if nature has been subpoenaed to give evidence of its own crisis; the political work that goes into the making of environmental problems falls away. Each of these places thus produces the "truth" of nature, a key aspect to the circulation of green governmentality.

However, there are two interesting discontinuities in this neat narrative. The first is Disney's focus on "un-truth." While Disney certainly utilizes scientific narratives in parts of the Animal Kingdom, in other parts it also fictionalizes them. Disney mixes the real and the make-believe, fact and whimsy, in their telling of the "truth" about nature. Rather than the presentation of dusty museum cases or dry lectures on the behavior of grizzly bears, Disney seduces with play and pleasure, taking elements of the scientific endeavor and blending them with popular culture to fashion a new kind of "truth," an environmental utopia based on fantasy, fact, and consumption. *An Inconvenient Truth* also flirts with fiction, but more fully utilizes another means to evince authority: autobiography and storytelling. Indeed, science and storytelling, climate and celebrity, are fused in the personage of Al Gore, where he and his life become the embodiment of the quest for truth through scientific knowledge.

Moreover, these sites do not only rely on science; they also all deploy the language and practice of commerce, another discourse of "truth," but one that is much more contested. Each place has a consumptive element that brings the market into their attempts for governing. Here Disney takes center stage. Indeed, it is difficult to determine where conservation ends and consumption begins at the Animal Kingdom, as the two are thoroughly enmeshed. Becoming a conservation hero means adding just one dollar to your other purchases; ecotourism is made nature's savior in Asia and Africa; tie-ins with blockbuster movies are used to narrate nature's crisis; and everywhere ideas of nature and its conservation are up for sale.

Of course, this probably comes as no surprise: Disney is one of the wealthiest corporations in the world for good reasons. That Disney takes aim at what is swiftly becoming a marketing goldmine—concern about the environment—and seeks to commodify it seems to fit comfortably within their business practice. But what *is* interesting is the result: this focus on conservation, no matter how it is deployed, cleanses Disney of the more repugnant elements of postindustrial capital. Disney appears to be a green corporation, a "good" citizen that not only seeks profit but also a better world, or, more properly, works to prove that the two are not mutually exclusive. In this way, Disney shores up what is becoming an increasingly pervasive narrative: certainly we have a problem with planetary nature, but we really just need to buy differently to forestall it.

The American Museum is not immune to this narrative. Although it frames the Hall of Biodiversity within the conventions appropriate to a learned institution, a backstory of responsible consumption is also present. But it is a more complicated approach than the one taken by Disney. The American Museum, in some ways, critiques the consumerism that has shaped the Western world, taking aim at the effects of deforestation and overfishing, agriculture, and urbanization. This analysis stands in tension with the fact that the Hall of Biodiversity was mounted in part by corporations like Monsanto and Bristol-Myers Squibb, which commodify and transform the very basis through which life exists. This might point to why corporations are never named as agents in the loss of biodiversity. So, consumption is named as a problem, and it is also proposed as a solution. Thus, when the American Museum turns to how the biodiversity crisis might be ameliorated, the citizen–consumer is once again recentered: travel and consume responsibly (and indeed the American Museum offers its own scientific ecotours), reduce energy demand, adopt a simpler lifestyle, appreciate biodiversity as a set of ecosystem services that can be valued at market price. The economics of conservation is made the answer to a biotic world under threat.

Similarly, *An Inconvenient Truth* appeals to the citizen–consumer in its solutions to the climate crisis. Viewers of the film are exhorted to green their consumption patterns, from buying a hybrid vehicle to purchasing compact fluorescent light bulbs. In recent years, Al Gore has modified his vision of potential solutions to the climate crisis to include a whole host of more fundamental (though technologically oriented) societal shifts in energy production and consumption, as explored in chapter 4. But part of the reason why *An Inconvenient Truth* retains its potency is the simplicity of its consumer-based solutions; like the other cases in *Governing the Wild*, it offers relatively apolitical mechanisms for political problems.

Finally, of course, the ecotour also functions by rendering nature into product; in this case, it offers its clients the experience of a nature they fear is fast disappearing. Indeed, akin to all the cases in this book, consumption equals action: a trip to the wilds of nature is recast as a form of activism. Preservation functions through commodification, and the ecotour puts a price tag on communion with nature that can never, in fact, be wild. So, the ecotour also hinges upon the visitor's ability to apprehend that nature's rescue from destruction turns upon how effectively it can be enrolled into the market.

In essence, two discourses of "truth" become one in these sites: science and commerce meet to good effect. Scientific discourse becomes the vehicle through which the environmental crisis can be named and assessed, and reliance on the market becomes the mechanism to solve it. Corporations are no longer the enemy of the environment, but its sponsor. Thus, while there are critiques of consumerism present in all four cases, they remain thin at best. Simultaneously, while "we" are named the agent of environmental destruction, "we" also have the ability to forestall it, most effectively through purchasing choices. Individuals are the central players here as the exhortation not to consume is cast aside in favor of an admonition to simply consume better. Each site examined shies away from any sustained critique of the broader structural forces that also make environmental problems, instead offering easier and necessarily limited engagements with why environmental issues emerge and how they can be addressed.

The Production of Experts

Each of these sites not only generates discourses of "truth," but also the various authorities necessary to speak it. The American Museum finds itself the frontrunner with regard to the production of unassailable experts, likely because it displays the dispassionate, guileless, and distanced mastery of science. Engaging in a modern-day version of Pratt's (1992) global project of classification, the scientists of the American Museum scour the world (and their own immense collections) to catalogue its species and (re)produce them for exhibition. Science, here, is made intelligible for the masses. Moreover, the museum has become a factory for the production of new authorities, with its graduate programs in comparative biology. The American Museum is a site where scientific expertise is made, honed, and deployed.

However, this image of the scientist is complicated by the museum's legacy of expeditionary collection. So, here we have the scientist as not only fact-seeking and objective, but also as daring and self-sacrificing. Recall the mounting of the Dzanga-Sangha diorama, where museum luminaries like Carl Akeley were invoked to describe the exploration of the rainforest of the Central African Republic in search of new specimens of biotic diversity. These are scientists who have not only mastered the laboratory, but also the field, eschewing comfort and courting danger in the pursuit of science.

The narrative of the intrepid scientist also finds a home at Disney's Animal Kingdom. The new ride at the Animal Kingdom, Expedition Everest, was "Imagineered" in part through the trek to Nepal undertaken by Disney and Conservation International to discover new species. The Animal

Kingdom also affords its guests the possibility of engaging in their own expedition, a fantastic journey back in time to rescue a dinosaur. Disney is elevated to higher purpose: not just theme park but also scientific agent. But Disney does not only rely on the expedition, fictive or otherwise, to profess its claim to expertise about nature. It also hosts a whole range of animal programs and research projects, as well as the Disney Wildlife Conservation Fund, all explored in chapter 2. It employs animal biologists, veterinarians, and keepers, it connects to other experts through networks like the Association of Zoos and Aquariums, and it established and expands upon its Environmentality Program. In each of these ways, Disney not only positions itself as the inheritor of a scientific project, but in so doing remakes itself as an expert in corporate citizenship. By invoking scientific authority and environmental commitment, Disney is purified of the aura of big business and reformulated into an amalgam of research institution, nongovernmental organization, and corporate innovator, all made safe, neat, and certainly nonconfrontational through a brand "we" trust.

The ecotour also relies on scientific experts to narrate the "truth" about the wilderness its clients seek to experience. In this case, the guide is the authoritative interpreter, educating the ecotourists and prescribing appropriate action for environmental preservation. Watching wolves gave birth to zoological descriptions, natural features made interesting geology lessons, examining trees led to explorations of ecosystem management. In this way, the guide of this trip through Yellowstone and Grand Teton not only performed his authority as an agent of science, but also worked to make his clients into budding scientists, and because of this, real environmentalists.

Like the guide, Al Gore, former vice president, celebrity, author, green entrepreneur, and climate activist is *the* expert of *An Inconvenient Truth;* indeed, he is the conduit through which all other expertise flows. Gore acts as a node of knowledge dissemination, engaging in the authorial practice of collecting and reassembling various forms of evidence about the climate crisis, repackaging them into a neat narrative that can be viewed in ninety-four minutes. Along with science, his autobiography lends him authoritative weight. It is through Gore's life story that movie goers come to know Al Gore as learned, yes, but authentic as well.

Each of these sites, then, offers us an ensemble of experts, variously positioned to help the citizen–consumer understand and experience nature. Without these experts, the knowledges produced at each of these sites would lack coherence, and the visitor might find herself rudderless in a sea of potential understandings. The profession of expertise allows the visitor or viewer to trust the knowledge generated as factual and hence important to acknowledge and act upon. And yet, this profession of expertise is never a fully completed endgame. It requires re-articulation and renegotiation. This is particularly evident when expertise is centered in a single figure, like Al Gore. As shown in chapter 4, there have been ongoing attempts to discredit his moral stance, from examinations of his excessive energy consumption to charges that he is positioned to become fabulously wealthy from the climate crisis. More generally, the expertise of the environmental sciences is under continuous pressure from various agents who seek to dismantle the consensus around climate change, as witnessed, for example, in "Climategate." And when corporations like Disney or the ecotour claim scientific or environmental expertise, criticism is swift and vociferous. As such, authority is not once and for all, but rather is a contested and contextual beast, relational all the way down. What experts in each of these cases do, then, is attempt to secure, albeit contingently, a toehold on the production of truth, a space to claim and reclaim solid ground in the quest for knowledge, power, and ultimately, the ability to tell truth. While never complete, what each of these cases show is that this claim is powerful, shaping, in many ways, how environmental issues are imagined.

Strategies of Regulation

What strategies do these experts effect? What registers do they work upon? Part of what these authorities do at these places is create ways of seeing—technologies of power that make the world intelligible. This intelligibility rests on vision to narrate nature's "truth." This is most evident with the ecotour, where sight is privileged as the appropriate sense through which one encounters wildness. Indeed, the ecotour seems a particularly apt expression of Urry's (2002) tourist gaze, where the signs that define wilderness are assembled in a visual grammar, made for consumption by eager tourists keen to experience the difference or singularity of a place and its animals. Seen through the LCD screen of a digital camera or the eyepiece of a high-powered scope, the spectacular natures of Yellowstone and charismatic beasts like wolves can be fixed in time and space. In doing so, the photographer/tourist can reassert mastery over the landscape while simultaneously dispelling the anxiety felt about threats the animals face. Not only the object of the gaze, but also the subject who performs it, are governed in this move, as places become postcards, animals become reprints, and tourists become photographers in search of the perfect picture to denote their trip. So Old Faithful, Cooke City, the Teton Mountains, as well as the bears, bison, wolves, elk, and other species encountered on this trip become representations of wilderness: their meanings and materialities more and more limited so as to conform to tourist expectations. These limited representations also travel, moving across the globe to demarcate what they are, and can be.

The gaze is also central to the work of Disney's Animal Kingdom; indeed, each attraction functions to organize the vision of the guest so that he or she understands Disney's view of nature and conservation. Recall how the Animal Kingdom's visitor is invited to observe Disney's efforts at conservation at Rafiki's Planet Watch; veterinary procedures quite literally perform the "authenticity" of Disney as an agent of environmental change. Moreover, the fauna and flora of the Animal Kingdom, both real and imagined, serve as plot devices: bumping along on the Kilimanjaro Safari ride, walking through the Pangani Forest Trail, sighting hidden animals on the Tree of Life, floating down the Kali River Rapids ride—all of these attractions work to tell the story of Disney's concern with nature and the appropriate ways to "save" it. Our vision is corralled, made to take notice of Disney's particular view. Nature and its narratives, as well as the places that house it (which, of course, are Disneyfied and removed from their biophysical and cultural surrounds) can be consumed in a daylong trip to the Animal Kingdom rather than a longer engagement with the animals, plants, and people that make up the places they represent. The Animal Kingdom functions, like the rest of Disney's universe, through spectacle. But it is a spectacle that instructs, providing ways of seeing that position environmental problems as issues of individual responsibility and economic opportunity, rather than complex, multiply situated, involving assemblages of human and nonhuman actors, and tied to capitalist modernity.

The Hall of Biodiversity at the American Museum also operates through the use of spectacle, but of a different kind. The "Spectrum of Life" and the Dzanga-Sangha diorama beget awe at the immensity of biodiversity, casting the gaze toward the achievements of the scientific endeavor in cataloguing nature's bounty. The "Spectrum of Life" renders nature's progress intelligible, mapping all life, and all possibilities of life, for the visual consumption of the museumgoer. The rainforest diorama ensconces viewers in the drama of environmental destruction, watching the movement from pristine to degraded in a ninety-foot-long replica. Vision in both of these cases is compressed, so that the viewer can understand both the value of biotic diversity and the challenges it faces in a glance. The technologies of vision deployed at the museum define nature for us, and in this move, they position scientists as the authorities whose knowledge can save it.

An Inconvenient Truth also deploys particular kinds of vision in its presentation of the climate crisis. The documentary uses images, statistics, simulations, temperature predications, diagrams, and graphs—in short, technologies of vision deployed through science—to map the state of the globe's nature and to narrate a compelling lesson of what happens when politicians and citizens do not listen to the wisdom of science. But spectacle in *An Inconvenient Truth* works not only through the kind of scientific mechanisms of display seen, for example, in the museum, but also through the demands of filmmaking. The movie entertains through its production of shock, elicitation of fear, generation of longing, and stimulation of guilt. It is cinematic and thus must rely on the sensory abilities of vision even more fully than the other cases of this study.

All four of these cases present visualizations of nature in their own distinct but related ways. They measure, document, and assess nature to produce "truth" effects about its situation. And these "truth" effects generate scripts for how nature *should* be read. The ecotour casts nature in the role of savior, the wellspring through which modernity can be redeemed, if only we act to preserve it. The theme park makes it into a phantasmagoria, a swirling collection of images, imaginings, and representations: fictive and factual beasts that work to narrate the story of nature under threat. The museum apportions nature into its proper categories, creating an orderly but endangered vision of the globe. The film makes nature into a plot device that renders threat both visible and, because of its visibility, palpable. At each of these places, the masterful eye is the apparatus necessary to apprehend nature, its problems, and its solutions. It is in the structuring of the gaze that we come to know nature as some particular things, and not others. In each case, we come to know it as scientific, disappearing, and ripe for salvation through science and the market.

However, these sites work not only through vision but also affect to govern the humans who visit them. Each of the attractions, practices, stories, and representations in these cases comes together through a range of registers, eliciting nostalgia and wonder, anxiety and hope, pleasure and desire, guilt and despair, longing and love. They enlist a variety of tropes to work on our affective selves; through these sites the visitor or viewer can access discourses that gain coherence from stories we already know: romantic wilderness, redemption through science and the market, the joys of right living, the fecundity of Southern bodies, the fear that nature will disappear before we can know it. By evoking our bodily senses and psychological desires, the tropes deployed by the American Museum, the Animal Kingdom, the ecotour, and *An Inconvenient Truth* offer the potential to predetermine our understanding, writing on our selves the path that nature's rescue should take.

The nonhumans (ghostly or living) that inhabit these places are subject to a different kind of regulation. Indeed, there is an attempt to render them *affectless,* without capacity, agency, or force. Because of this, the nonhuman is subject to another form of governing at these sites: disciplinary regulation. I mean this not only in the literal sense of imprisoning animals, by way of the chain-link fences at Disney, the boundaries of the national parks, or the display cases of the American Museum, but also in their very possibility to be other than what they are scripted as. So, the museum generates biotic diversity as wondrous, yes, but also as manageable through the correct application of scientific knowledge. Disney produces animals as a wellspring of love, but also as beings whose main purpose is to entertain and educate humans. The ecotour positions wildlife as a tool for self-redemption—to gaze upon animals is to do work on our own souls. The film, when it does reference animal life, depicts it only as endangered, subject to the will, nefarious or salvational, of human beings. In each case, then, animals are not themselves (whatever that might mean), but rather stand in for something else: they are rendered into simulacra to teach us about human motivation and action. They are enrolled into biopolitical projects that measure and count them, catalogue and document them, ascribe value and meaning to them, but never allow for the possibility that the nonhuman will always be more than we can know, will always be outside of our representations of it. Thus, they are rendered mute, and humans (of a particular kind) are made, to paraphrase Haraway (1992), into their ventriloquists. The encounter between human and nonhuman at these sites operates at strict behest of the authorities that assemble the animals and plants to tell a narrative of "truth." Of course, this is an illusion. Animals (including humans) as well as plants, bacteria, technologies, and the host of other organisms that are lumped together under the shifting and necessarily imprecise category of nonhuman *do* act in the world, forming both mundane and novel combinations that make up our world. Indeed, I will later suggest that it is the recognition of these assemblages that might hold the key to a different kind of ethics, a potential way out of green governmentality. However, for now, it is important to note that at each of the sites there is no room for the agency of the nonhuman, nor an ability to step outside of the boundary-making process that separates "us" from "them." Rather nature is enrolled in a truth game, where it serves as the vehicle to tell us about ourselves.

Subject Formation

Also generated at each of these places is a very particular subjectivity that can be taken on by its visitors, which I have been referencing throughout this conclusion: the environmental citizen as consuming subject. Various practices of self-fashioning exist at each site. At the American Museum, the now-learned subject (schooled in the scientific basis of life) comes to the Solutions wall ready for action: to support biodiversity research, reduce our ecological footprints, and encourage corporations to produce green products. At the Animal Kingdom, we need only buy something to become a "conservation hero": through donating a dollar, the visitor can feel secure in their efforts to save the earth. The ecotour is itself the vehicle through which our consumer choices make environmental change: by choosing to go on the ecotour to begin with, the visitors have marked themselves as people who care about nature (once fully equipped with the right gear). And the documentary works to scare the crap out of its viewers so that sustainable consumption is not just a practice, but a duty. The technologies on offer here work on both body and mind, attempting to condition how we can act upon the environment, and conceive of this action. And all of these efforts at subject-making are done within the rubric of consumption.

Of course, the seductiveness particular to the environmentalist project of self-regulation is that it can come with the possibility of claiming an identity of innocence and resistance to authority. As Foucault (1990) has remarked, there is pleasure in "speaking truth to power." In naming that which oppresses, in testifying about its domination, in critiquing practice, there is an attempt to position oneself anterior to, or outside of, power. In any incarnation, this would be a fiction. This is doubly the case when we turn our attention to the subjectivities generated in these places, where the environmental citizen *cum* consuming subject reinforces those same structures that lead to the specific forms of biophysical degradation. That is, in part, what makes the adoption of these subjectivities so easy: we need only tweak that which we already do, consume better, consume less, but for heaven's sake, don't stop consuming. Of course, these subjectivities are not freely accessible: they work through preexisting tropes, particularly of race and class. As such, these subjectivities remain firmly ensconced in the concatenation of power, "truth," politics, and profit that define our world.

So, I want to say two things at once. Indeed, I want to put forward two ideas that appear to be in conflict with one another—a story of both difference and similarity. First, divergence. The task of each of the chapters in this book has been to suggest that there are multiple ways to enact a framework of governance, signaling that green governmentality is multidimensional rather than monolithic, and because of this, more difficult to get a theoretical and political grasp on. The practices, discourses, and technologies at each site provide sometimes subtle, sometimes striking variations in how notions of the nonhuman are made and circulated. Scientific assessment of global environmental threat, corporate biopolitical practices of conservation and consumption, aesthetic regulation of animals and people, and moral authority to speak truth are the respective watchwords here. These heuristics allow for delineation between multiple mechanisms of government, different ways through which knowledge is made, circulated, and taken up.

And yet, despite the multiple ways green governmentality is enacted in these sites, the cases do show interesting areas of congruence. While the museum, the theme parks, the ecotour, and the documentary offer different experiences of nature for different kinds of subjects, they do sometimes perform quite similar work. On the surface, these cases might be read as quite dissimilar: the museum as storehouse of scientific knowledge, the theme park as playground for families, the ecotour as a journey to see wildlife in its natural habitat, the film as a visual and political narrative of nature under threat. But it is not how they appear but what they produce that I am interested in. They render "truths" visible and intelligible, they generate experts to define and explain these "truths," they assemble technologies of power and cultivate strategies for intervention, they articulate biopolitical projects, they foster particular subjectivities, each in different ways. In doing so, however, they proffer similar narratives and scripts, they provide us with a lexicon through which to apprehend the world, and they tell us what nature is, how it is under threat, and the appropriate solutions to its crisis. These are the hallmarks of green governmentality. They are shopping malls and playgrounds, schools and prisons, zoos and research labs. Each contains so many narrative and representational elements of the other that it becomes difficult to tell them apart. They are also all sites where nature goes to die, if not literally then in our imaginations. What all these sites do is narrow our ability to imagine nature as something other than what the "truth" tells us. So, these are also dangerous places, filled with blind spots, violence, erasure, and closure. To be otherwise is strictly prohibited.

The production of knowledge, practice, and subject–positions at these places might not matter so much if they occurred in isolation. But what is crucial here is that these stories do not stand alone; they exist within a web of allied discourses and practices, making them nodes in an emergent regime of truth. Indeed, I would argue that the stories told at the American Museum, the Animal Kingdom, on the ecotour, and through *An Inconvenient Truth* find their home among a complex panoply of discourses, practices, and technologies that work to position a particular kind of environmentalism as the saving grace of the world's problems. These sites join an ensemble of corporations, governments, media, scientists, nongovernmental organizations, institutions of culture and recreation, universities—all known as experts—to produce grids of intelligibility through which these crises can be understood. They provide us with the intellectual grammar to apprehend the world. They give us easy answers to incorporate seamlessly into our own lives, with very little disruption. What is missing here is the very nonhuman nature upon which these discourses and practices rely. Indeed, different knowledges about nature, both human and nonhuman, are shunted aside and the possibilities for alternate discourses, practices, and subjects are narrowed as the regime of truth gains strength. But that does not mean that other knowledges have ceased to exist—they proliferate, multiply, and insist on airtime. So while the story provided through this form of green governmentality is a compelling one, it, like all other discourses, is vulnerable to those subjugated knowledges it has placed on the margins.

Where Do We Go from Here?

The pertinent political question, then, is how do we unwrite or rewrite these stories that have been so well crafted? How can there be space outside of this green governmentality? I am sad to say that I provide no coherent answers here, no roadmap for change. Indeed, I think such an effort would be in vain; there cannot be a single way to approach this question. But what I hope to do in closing this book is to make a small gesture toward a *relational ethics,* an opening for the possibility of radical difference that might allow for an imagining of other kinds of encounters between the human and nonhuman.

Perhaps what is called for is Foucault's "insurrection of subjugated knowledges," which have been "buried" or "disqualified," "naïve knowledges, located low down on the hierarchy, beneath the required level of cognition or scientificity" (1980, 81–82). Thus, it is not so much a project of inventing new grammars of intelligibility (for this would seem to reinscribe the same totalizing structures it is important to resist), but rather recognizing that alternate ways of being have always already been there, they have simply been occluded from view by "the tyranny of globalising discourses" (83). However, once again, Foucault's "species chauvinism" gets in the way. What he never explores is an outsideness to discourse, one that might be afforded if we reconceptualize nature as actant rather than passive vessel, as co-constituting "truth" rather than simply subject to it. The question, then, is what might this reconceptualization generate?

To think through this question, I take the hand of those who (by self-designation or otherwise) find themselves under the umbrellas of posthumanism and actor-network theory, not because they are the only scholars who think through such concerns, but rather, in different ways, relentlessly interrogate human/nonhuman/inhuman assemblages and insist on contingency that is vital, in my view, to the possibilities of being otherwise. Because, what I want to suggest is that it is in the recognition of a relationship with and responsibility to nature that we might be able to find a new way of thinking about both the human and nonhuman in ways that more clearly reflect the workings of a complex world but also offer a sense of justice that, in my view, is missing from the cases outlined in *Governing the Wild.*

Writing as though she might be remarking on the case studies of this book, Donna Haraway remarks in "The Promises of Monsters":

> Efforts to travel into "nature" become tourist excursions that
> remind the voyager of the price of such displacements one pays to
> see fun-house reflections of oneself. Efforts to preserve "nature"
> in parks remain fatally troubled by the ineradicable mark of the
> founding expulsion of those who used to live there, not as innocents
> in a garden, but as people for whom the categories of nature and

culture were not the salient ones. Expensive projects to collect "nature's" diversity and bank it seem to produce debased coin, impoverished seed, and dusty relics. (1992, 296)

These kinds of natures, she argues, lead only to "reification and possession" (296). Moreover, they lead to the remaking of all things (organisms, people, technologies, ideas) into a replica of the commodity form: "The preoccupation with productionism that has characterized so much parochial Western discourse and practice seems to have hypertrophied into something quite marvelous: the whole world is remade in the image of commodity production. . . . Hyper-productionism refuses the witty agency of all the actors but One; that is a dangerous strategy for everybody" (297). Haraway argues against this, proposing a politics of articulation rather than representation. If representation denies complexity, remaking nature into the image of silent object that always must be spoken for (either in the scripts of saving or destroying), then articulation requires something wholly different. She invokes the figure of Trinh Minh-ha's "inappropriate/d other" as an actor who is neither self nor other, but inhabits and haunts an in-between space. Trinh Minh-ha is referring to both the instability of identity, and thus the fiction in claiming positions of self or other, but also those hybrid bodies that do not belong, that refuse to be narrated with this dichotomy. In this sense, the "inappropriate/d other" is "someone whom you cannot appropriate, and . . . someone who is inappropriate. Not quite other, not quite the same" (Trinh Minh-ha 1998). Haraway suggests this figure can also operate as a foil to the fixing of nature and culture into distinct categories, which offers an opening to recognize that nature is both of and more than its representations by humans. For Haraway, then, following Trinh Minh-ha, the "inappropriate/d other" violates such boundaries: it does not "fit in the taxon" (1992, 299). The work of representation serves to distance "us" from "them," drawing firm lines around nature and culture, subject, and object. Conversely, articulation insists that we are bound to each other in relationships of differential power, that we recognize that the project of separating nature from culture has always been an illusion, but one with real material and discursive effects.

The figure of the fractal, drawn from mathematics but deployed by Marilyn Strathern (2004) in anthropology and John Law (1999) in geography is, I think, a useful metaphor to ponder this complexity, this always already enmeshment. A fractal is an object of infinitely complex composition, which at smaller and smaller scales reveals more and more intricacy. A fractal has two important characteristics. The first is self-similarity, where patterns repeat themselves, but never in exactly the same way. Each part of a snowflake, for example, bears a resemblance to the whole snowflake in which it is nested, but each is infinitely different. The second characteristic, counterintuitively (or at least for my nonmathematical mind) is that a fractal curve inhabits more than one dimension, but less than two. A fractal object, then, sits between dimensions—it does not occupy a whole dimension but rather seeps between multiple dimensions. And so, there is a necessary *indeterminacy* to the fractal. The most famous application of this is by Benoit Mandelbrot, who used the example of the British coastline to demonstrate how fractals work. An application of Euclidean geometry to the coastline would approximate, simplify, or idealize, making the sand and rocks fit into strict categories of dimensions. But as Mandelbrot showed, at smaller and smaller scales these measurements change. A straight line is never just a straight line; rather, a coastline (or a tree, leaf, mountain, icicle, etc.) is made up infinite possibilities for difference.

So, what might the fractal offer us for a reconceptualization of nonhuman/inhuman/human imbrications? First is the fractal's inherent complexity; its shape is similar to its larger iteration, but it is broken, irregular, and fragmented. It *grows*, occupying multiple dimensions, but never completely. Thus, the image of the fractal necessitates an attention to infinite possible combinations, recombinations, articulations, and re-articulations, demonstrating that the harder we look (feel/encounter/sense), the more there is to see (feel/encounter/sense). But it also is a figure that unsettles taken-for-granted assumptions about how the world works, like Trinh Minh-ha's "in-appropriated other." As Law suggests, the fractal is "difficult to think because it defies the simplicities of the single—but also the corresponding simplicities of pluralism of laissez faire, of a single universe inhabited by separate objects. So the thinking is difficult—no, it is not transparent—precisely because it *cannot* be summed up and reduced to a point, rendered conformable and docile" (1999, 12, emphasis in original). What the fractal does for the imagining of and encounter of these relationships is signal two things at once, both of which I think are key to disquieting the presumed order and certainty that green governmentality provides. It suggests an instability to our relationships with the natural world and the categories that have been used to describe it. A fractal cannot be fixed, it is complexity all the way down, with infinite fragmentations, enmeshments, and combinations. But I also think what is important here is an element of incomprehensibility—that these relationships will always be more than we can grasp, intellectually or affectively, but that doesn't mean we shouldn't try. So there is an uncertainty here—a tentativeness—that I think is vital in rejecting the kinds of simple causes or solutions to environmental problems that the cases of this book conjure, and instead considers how humans and nonhumans have always made each other up.

However, as Haraway contends in her more recent book, *When Species Meet* (2007), that recognition of the complexities inherent in relationships between the nonhuman and human that the metaphor of the fractal so successfully describes isn't sufficient to developing an ethic of relationality. Or more properly, the political work can't stop at recognition alone. To combat the human exceptionalism that has largely characterized ideas about the nonhuman and their place in the world, she invokes the notion of "companion species"—of "coshapings all the way down, in all sorts of temporalities and corporealities" (2007, 164). Drawing on Mary Louise Pratt, Haraway suggests that these coshapings emerge in *contact zones,* a notion that emphasizes the relational character of the encounter:

> [I]t's not about who "has" hermeneutic agency, as if it were a nominal substance instead of a verbal unfolding. Insofar as I (and my machines) use an animal, I am used by an animal (with its attendant machine). I must adapt to the specific animals even as I work for years to learn to induce them to adapt to me and my artifacts in particular kinds of knowledge projects. Specific sorts of animals in specific ecologies and histories make me adapt to them even as their life doings become the meaning-making generator of my work. If those animals are wearing something of my making, our mutual but unidentical coadaptation will be different. The animals, humans, and machines are all enmeshed in hermeneutic labor (and play) by the material-semiotic requirements of getting on together in specific lifeworlds. They touch; therefore they are. (164)

An honest realization of co-constitution and encounters *as becoming* necessitates a particular kind of politics for Haraway: one of response and respect, or "response-ability," which is premised on a mutual flourishing. Thinking with the notion of medical testing, Haraway explores the ethics of making some animals killable. She accepts that animals used for such experiments are significantly unfree and subject to all manner of painful and life-shortening interventions. But also that the politics needs to be more complicated than simply that assertion. If the nonhuman and human are co-made all the way down, as she suggests, then the practice of medical testing is an important one, but also requires responsibility, not only to lessen pain, but also to share suffering as asymmetrical partners (72). This approach acknowledges that all organisms and technologies in such experiments are participants and, as such, there must be both imagination and engagement to think about how an ethics can proceed. And so, as Cary Wolfe suggests, what is required is "an increase in the vigilance, responsibility, and humility that accompany living in a world so newly, and differently, inhabited" (2010, 47).

And so, for me at least, a reimagining of the sites of this book, and of the practices and discourses they house more generally, requires the kind of relational ethics that Haraway proposes to render them more palpable to questions of justice for humans and nonhumans alike. Let me think this through with one example—the ecotour. As I have described, the ecotour operated through a kind of prescriptive aesthetics, a rendering of the eye as the primary sense to comprehend majestic nature and charismatic megafauna. But what if this were not the case? What if rather than watcher and watched—subject and object—the intersection of human and nonhuman was reframed as an *encounter?* This is not to suggest that there wouldn't be power relations laden in such a meeting, but rather that there would be recognition of both the human *and* the nonhuman as participants in this dance. Of course, the animals that the ecotourists travel to Yellowstone to see are unfree in some sense, rooted as they are in place by not only by the boundaries of the national park but also the human practices that have destroyed their habitats. However, understanding the meeting as an encounter would not deny all the practices that have shaped the lives of both nonhumans and humans alike, such that we arrive at this place and this moment. It wouldn't separate us from them. Thinking back to the fractal, with its infinite complexity, those of us on the trip would be required not to make easy distinctions, but to see how our lives are bound to the animals we've come to look upon. And indeed, it would require more than simply looking upon; it would necessitate engagement, response, interaction. Taking the example of the radio-collared wolves, the ecotourists could take into account the economic, scientific, political, historical, technological, social, biophysical, and cultural practices and encounters that led to these "Canadian" wolves finding a home in land emptied of its human and nonhuman inhabitants so that it could be made into wilderness for us to visit. Indeed, it would oblige the ecotourist to leave aside the illusion of wilderness altogether. This would offer an opening to the fact that animals look back, that they respond to our presence, that they also encounter us as we encounter them. The trip, then, would not be about cataloguing images of flora and fauna (or images of images) in some race to assemble markers of wildness. Instead, it would be an opportunity to know something about the animals rather than our reactions to and visions of them. And in this encounter we might find a kinship, where, as Haraway notes, everyone has a face. In this knowledge there would be responsibility, not just to document what is thought of as the last of a disappearing and fictional wilderness, but rather to make the lives of these animals, hampered as they have been in their co-evolution with humans, more attuned to notions of flourishing.

While this effort at reimagining might seem more science fiction than actual possibility, I hope this is not the case. For I think it is imperative to get these things right, not only to try to work against the limitations and self-certainty that green governmentality seems to offer, but to try and deal with the very real environmental justice issues like climate change and habitat destruction, communities made toxic by industrial capitalism and factory farming, all of which make up our shared and uneven world. Indeed, what this reimagining offers is the notion that it is necessary to refigure our interactions with the nonhuman as a collective of multiple actors and unlikely alliances, instead of one that signifies and one that receives such signification. Of course, the human and nonhuman are made up of but are also more than the discourse that describes them. If we accept, as Stacy Alaimo suggests, a reconfigured understanding such that nature is not "a distant refuge from culture but an ally that is always as close as our own skin" (2000, 178), each of these sites is but a reflection of our own practices of signification, far removed from the human/nonhuman assemblages that constitute the world. Understanding this relational ethics insists that we recognize Latour's (1993) quasi-objects, the entangled nature–cultures that complicate our view of the world, but nevertheless make it more fulsome and present opportunities for just encounters. This stance would not be akin to animal rights, for indeed, questions around the extension of rights and the capacity for sentience seem to be the wrong ones to ask. Instead, we must stay with the complexity. If we go down this path, we risk uncertainty. The green governmentality explored in this book has been about defining and fixing the appropriate relationships between the human and nonhuman through the lens of science, the practice of biopolitics, the making of subjectivity, and the dominion of the commodity form. Among other things, these efforts provide conceptual safety; they are familiar and comfortable narratives to describe our world. A relational ethics would unsettle and disrupt this familiarity, forcing the recognition of the multiple actors, both human and nonhuman, that form our bodies, ourselves, and our world. But it offers, I think, a way to find spaces of otherwise, outside of green governmentality. And that is something we desperately need.

Acknowledgments

It has become commonplace that acknowledgment pages begin with the comment that one never writes a book alone. Although it risks cliché status, it nevertheless remains true. There are teams of people who deserve many thanks for the part they played in the production of this book, from the editorial staff at the University of Minnesota to the colleagues, friends, and family who assisted with the process.

First, without my editor, Pieter Martin, and the University of Minnesota Press, this book would never have come to be. Pieter's faith in the project, which was made evident through his gentle encouragement, suggestions for revision, and numerous cups of coffee, assisted a junior scholar in finding her feet. Kristian Tvedten also provided an invaluable editorial assistance, steering me through my computer-phobic thicket to generate the images that accompany this book. Simon Dalby provided keen insight as one of the reviewers of the manuscript. The other reviewer, who remains anonymous, was equally instructive providing trenchant comments that improved the final manuscript significantly. Sue Breckenridge served as copyeditor for the book. Her careful reading and insightful commentary greatly improved the readability of the book. Finally, Denise Carlson was the indexer for the book, expertly performing this important work as I prepared for the birth of my first child.

This book, as some do, started life as a dissertation. Although in many ways it is quite different from its initial articulation, I am thankful that my doctoral committee insisted I "write it like a book," rather than as a dissertation. I am grateful for their sagacious advice in this respect, as well as their guidance throughout the project. Thus, special thanks are due to Gerda R. Wekerle, Liette Gilbert, and Steven Flusty. I would also like to extend much gratitude to the many mentors, colleagues, and friends who have helped along the way, from reading drafts to words of cheer or consolation. In particular, I would like to recognize Ellen Arnold, Pablo Bose,

Bruce Braun, Darren Dias, Bruce Erickson, Leesa Fawcett, Roger Keil, Tim Luke, Deborah McPhail, and Jocelyn Thorpe.

This project was also initially supported through a Social Science and Humanities Council of Canada doctoral fellowship as well as an Ontario Graduate Scholarship, both of which allowed me to focus on writing, for which I am certainly grateful.

And, of course, without one's family and loved ones, such ventures are next to impossible. My father, Paul Rutherford, to whom this book is dedicated, was particularly instrumental. He took an active role in helping me conceptualize the book. Over numerous lunches, dinners, and weekends, he talked me through the process of writing and rewriting, facilitating my own vision and voice. Without his intellectual contribution as well as emotional support, this book would be much less than it is. My mother, Gail Brankston, and sister, Jennifer Rutherford, are both my impassioned champions; one does benefit from people thinking you are brilliant, even if it's less than true. My stepfather, Craig Brankston, generously produced the wonderful maps included in the book. My stepmother, Margarita Orszag, has also provided a grounding influence, and made it so that I did not take myself too seriously. My grandparents, Bill and Blanche Rutherford, offered yet another layer of assistance. They both died before the publication of this book, but their seemingly endless capacity for support remains with me. Finally, I would be remiss if I didn't thank my partner, Darcy Bonner. His intellectual questioning and constant enthusiasm for the project have been invaluable. He unwaveringly believes in my ability to do just about anything and often works as my own personal cheering section. A writer of a different kind, he challenges me to think and write more clearly.

Notes

Introduction

1. Postcolonial scholars such as Timothy Mitchell (2002) and David Scott (1999) have been extremely helpful in sketching out the ways the governmentality works through the making and enforcement of categories of difference. Also interesting are the works of Matthew Hannah (2000) and of Michael P. Brown and Paul Boyle (2000) that both look at the construction of censuses as a means to define populations (or erase them) based on race and sexuality, respectively. These studies, among others, show how governmentality normalizes, makes natural, and, through this process, makes visible certain kinds of citizens or subjects. But they also demonstrate that, simultaneously, governmentality can render other subjects as unnatural and abnormal, and occlude them from view. So my critique is not that no work has been done on this very important issue, but rather that the ways in which difference functions through governmentality should be a central aspect of any study because technologies of rule do not meet their targets with the same effects for each and all.

2. See, for example, Agrawal 2005; Bäckstrand and Lovbrand 2006; Birkenholtz 2009; Darier 1996; and S. Rutherford 2007.

1. Ordering Nature at the American Museum of Natural History

1. By *episteme*, Foucault (2004) refers to the conditions of possibility for the production of knowledge at a given point in time.

2. Cladistics is a taxonomic approach that organizes beings in relation to evolutionary ancestry.

3. I will confine my discussion of natural history museums to the United States in order to provide the historical background necessary for my examination of the American Museum of Natural History. However, the work of museums scholars like Hooper-Greenhill (2001) and Bennett (1995, 2005) indicates that many of the same themes are present in natural history museums in Europe. In an interesting counterpoint to this analysis, Susan Sheets-Pyenson's (1988) book, *Cathedrals of Science: The Development of Colonial Natural History Museums during the Late Nineteenth Century,* uses the examples of museums in Canada, Australia,

and Argentina to explore how these narratives of progress moved and shifted when transposed to the colonies, offering, in some sense, a more grounded exposition of local cultures while at the same time relying on metropolitan centers for institutional support.

4. I will argue below, however, that the American Museum has retained this role. In contrast to Conn's arguments, the American Museum, if perhaps an isolated case, has maintained its function as a knowledge producer, often in concert with universities. So while not the focal point of scientific research, museums may still retain power in the production of knowledge. Their business is not just limited to dissemination.

5. This, of course, is not particular to the American Museum. Indeed, as Schiebinger (2004) notes in her book *Plants and Empire,* many of the botanical expeditions that early naturalists embarked upon were in the service of colonial expansion and followed trade routes. Borrowing a phrase from David Mackay (1996), Schiebinger names these scientists "agents of empire."

6. In fact, the American Museum, in concert with the Environmental Defense Fund, mounted a traveling exhibition called "Global Warming: Understanding the Forecast" in 1992. In some ways, it seems that this was a precursor to the Hall of Biodiversity and provided some lessons about how to present information about environmental issues. For example, before mounting the exhibition, surveys of potential visitors indicated that they would like to have access to information about what they could do about global warming (Falk and Dierking 1992). This focus on individual action has been translated to the Hall of Biodiversity, where one whole section of the exhibit deals with possibilities for solving the biodiversity crisis. "Global Warming," however, was not a permanent exhibition, and so it is correct to list the Hall of Biodiversity as the first permanent issue-based exhibition for the museum.

7. In fact, there is a substantial body of literature that indicates that, when entering a museum exhibit, visitors tend to turn to the right rather than the left between 70 and 80 percent of the time (see, for example, Bitgood 1995; Davey and Henzi 2004). As such, why the curators of the Hall of Biodiversity decided to place the introductory video to the left of the entrance remains a mystery.

8. I inquired in an interview as to why there were no humans represented on this massive tree of life, given that the rest of the exhibit emphasizes humanity as part of biodiversity. The main curator of the exhibit indicated that he wished to see humans in the "Spectrum of Life" but lost the debate with his fellow curators. So while people are situated as both the cause of environmental problems and the triumphant savior, they don't appear in its complex biopolitical mapping. Again separating nature and culture, nature is pictured as belonging solely to the nonhuman.

9. It is worth noting, however, that the Center for Biodiversity and Conservation (established in 1993) does much educational and scientific work in different

places around the globe but does not focus on only the hotspots of biodiversity but rather works in areas where there are local partners and in sites where they can extend their institutional expertise and interest in small mammals, invertebrates, fish, and the like (anonymous interview 2, 2006). The center partners with governments, nongovernmental organizations, research and educational institutions, and community groups to conduct research on biodiversity and educate the public about threats to it.

2. Disney's Animal Kingdom

1. For a comprehensive, though now dated, discussion of Disney's holdings, see Wasko 2001.

2. There has been a lot of controversy over the filming of *White Wilderness,* which may give clues to Disney's (at least early) view of nature. Disney imported lemmings from Manitoba to Alberta and then set about filming them in such a way that it seemed as though they were diving off a cliff to commit mass suicide (Sagi 1992). It has subsequently been proven that lemmings do not behave in this way; however, millions of people grew up (myself included) thinking that they did. Importantly, what this film did was manufacture drama in what might otherwise have been a dull nature film (see Mittman 2003). This kind of cinematic license is also central to the experience of the Disney theme parks (see Bryman 1995). In addition, what *White Wilderness* also did was script nature's truth, which was only years later recognized and accepted as false.

3. Audio animatronics are the human and animal robotic devices created by Walt Disney Imagineering that are used as part of their attractions. They move, make noise, and sometimes talk. Some famous examples are found at the Pirates of the Caribbean, Jungle Cruise, and It's a Small World attractions. The newest audio animatronic can be found at the Animal Kingdom as part of their new Expedition Everest roller coaster.

4. Rafiki's Planet Watch in some of the older Disney guidebooks is listed as a part of Africa (cf. Birnbaum 2005). However, the recent renovation and update of the area has promoted Rafiki's Planet Watch to the status of its own "land."

5. The Pocahontas myth, or that of the "good Indian woman," has special resonance in the United States and helped shape a burgeoning nation and legitimate colonial expansion (Faery 1999; Tilton 1994). The film retells the popular yet fictionalized story of Pocahontas, a woman of the Powhatan Nation who saved John Smith from execution at the hands of her father and developed a romantic relationship with him.

The Powhatan Renape Nation has decried the historical inaccuracies that were advanced in this Disney film. In particular, they note John Smith's account of his rescue by Pocahontas (she would have been ten at the time) as well as the relationship between them, were likely fabricated (Powhatan Renape Nation, n.d.). In reality, Pocahontas was captured in 1612 and was held hostage for one year by the English in Jamestown. She agreed to marry a man named John Rolfe as a condition of her release, was taken to England, and died in 1617.

6. Interestingly, toward the end of this show there was a moment of U.S. nationalism. The host, who spent most of the show talking about threatened species and habitats, used the American bald eagle as an example of an animal that has resurged, particularly in captivity. The audience clapped and ahhed, respectfully marking their national symbol. The host noted, "This is Hope. Hope is a living symbol of one of the greatest conservation success stories of all time." The notion of national natures is revisited in chapter 3, on ecotourism in American national parks, but it is interesting to note that a piece of American nature has been saved through what might be read as environmental foresight and conservation action.

7. Jane Goodall has connections to the Animal Kingdom in addition to her identification as an eco-hero. Her foundation has received funds from the Disney Wildlife Conservation Fund, and she attended the opening of the Animal Kingdom. *Amusement Business* magazine reported, "Dr. Jane Goodall thinks there is nothing better than the 'entire experience' Disney Animal Kingdom offers" (O'Brien 1998, 1). High praise indeed from one of the best-known scientists in the world. These kinds of endorsements lend legitimacy to Disney's efforts to remake itself as a scientific agent.

8. In an interesting contradiction, and one that runs throughout the Animal Kingdom, while we can all save the world, some of us, like the veterinarian, the research scientist, and Disney, can do it better. Here we have the privileging of particular experts.

3. Wolves, Bison, and Bears, Oh My!

1. The International Ecotourism Society indicates that the proportion of ecotourists with postsecondary education is falling. They surmise that this means that ecotourism is going more "mainstream."

2. I suspect the use of the term "members" here does not denote membership in a conservation organization (although intentionally or unintentionally, that is how it appears) but rather membership in the "Habitat Club," a customer loyalty program that offers discounts on trips, gear, and a range of "conservation benefits" that emerge from multiple trips. For example, if you embark on two trips, Natural Habitat "will plant a tree in your honor through the Conservation Fund's Go Zero program." Your fifth trip generates a "symbolic animal adoption" through the World Wildlife Fund. On your tenth trip and beyond, Natural Habitat will "pay to offset the emissions from your flight so together we can help strike a blow against global warming" (Natural Habitat, n.d.2). Like Disney's conservation hero

pin, your purchase of other items renders you the status of environmentalist through consumption.

3. In addition, Natural Habitat was listed as the top tour provider on the *Condé Nast Traveler* magazine's 2006 Green List.

4. This article recounts Tierney's preparation for, and experience of, his trip to Ellesmere Island in the Canadian Arctic to rehearse the miseries of Starvation Camp, which, in 1883–84, was the site of a U.S. Army polar expedition that became stranded for eight months, forcing them to engage in cannibalism.

5. Along these lines, Alice Wondrak Biel (2006) in her book *Do (Not) Feed the Bears: The Fitful History of Wildlife and Tourists in Yellowstone,* charts how the National Parks Service's "Leopold Report" of 1963 moved the management of national parks from one focusing on the aesthetic pleasure of tourists to an attempt to recapture precontact nature. "Restoring the primitive scene, according to the report, would require planting and removing trees along roadsides, obscuring all 'observable artificiality,' creating simulated buffalo wallows in order to spur native plant growth, and reintroducing certain wildlife species in order to 'enhance the mood of wild America.' Above all, the NPS should accomplish these transformations 'invisibly,' meaning that the signs and traces of management efforts should be concealed from the public" (88). As Adolph Murie (qtd. in Wondrak Biel 2006) opined at the time, this is hardly a return to natural wilderness but rather represents more intensified attempts at regulation.

6. This is made even more apparent with the use of digital cameras, where the photographer can preview their image and delete those that do not meet the standard of natural beauty defined both by the photographer and the stockpile of other images of the same place.

7. Indeed, almost every day of the ecotour provided an opportunity to buy things. A trip to the National Museum of Wildlife Art in Jackson ended, predictably, in the gift shop. Our trip to see the nature photographer in Cooke City was designed so that we could purchase his photographs. We stopped at the myriad gift stores in Mammoth Springs, Old Faithful, and the Grant Village Visitor Centers. For the most part, the ecotourists' purchases were small: key chains, T-shirts, postcards, and books.

4. Science and Storytelling

1. Cockburn and St. Clair offer a cynical and in my view unfair depiction of this event in Gore's life, arguing that he bears some responsibility for letting go of his son's hand, was absent in the recovery process because he was writing *Earth in the Balance,* and subsequently he used his son's accident, along with his sister's death, to make political hay (2000, 135–40).

2. To put this in some degree of perspective, on July 5, 2010 (the number of

hits changes daily), *Twilight: Eclipse* listed 12,800,000 hits. Michael Moore's *Capitalism: A Love Story*, perhaps a more comparable film, has 7,350,000.

3. I mean this in the Latourian (1993) sense of the term.

4. An interesting exception here is the notion of "Creation Care." This has particularly been taken up by evangelical groups to assert the responsibility to care for the earth as a gift from God.

5. TED is a nonprofit organization that brings together leaders, academics, activists, and public intellectuals to discuss issues of technology, entertainment, and design. See http://www.ted.com.

6. The gendered pronoun here is intentional, as Foucault argues that the truth-teller in ancient Greece is necessarily male, a citizen imbued with the capacity to speak.

Bibliography

Adams, W. M., and D. Hulme. 2001. "If Community Conservation Is the Answer, What Is the Question?" *Oryx* 35, no. 3:193–200.

Agrawal, Arun. 2005. *Environmentality: Technologies of Government and the Making of Subjects.* Durham: Duke University Press.

Alaimo, Stacy. 2001. *Undomesticated Ground: Recasting Nature as Feminist Space.* Ithaca: Cornell University Press.

AMNH (American Museum of Natural History). 1998a. "National Survey Reveals Biodiversity Crisis—Scientific Experts Believe We Are in Midst of Fastest Extinction in Earth's History." AMNH. Accessed May 27, 2011. http://www.amnh.org/museum/press/feature/biofact.html.

———. 1998b. "American Museum of Natural History Unveils 11,000-Square-Foot Permanent Hall Devoted to Biodiversity." AMNH. Accessed May 27, 2011. http://www.amnh.org/museum/press/feature/biohall.html.

———. n.d.1. "Hall of Biodiversity." AMNH. Accessed May 27, 2011. http://www.amnh.org/exhibitions/permanent/bio/.

———. n.d.2. "Science at the Museum: Exploration and Discovery." AMNH. Accessed May 27, 2011. http://www.amnh.org/science/articles/science.php.

———. n.d.3. "Timeline: The History of the American Museum of Natural History." AMNH. Accessed May 27, 2011. http://www.amnh.org/museum/history/.

Anderson, Kay. 1995. "Culture and Nature at the Adelaide Zoo: At the 'Frontiers' of Human Geography." *Transactions of the Institute of British Geographers* 20, no. 3:275-94.

Bäckstrand, K., and W. Lovbrand. 2006. "Planting Trees to Mitigate Climate Change: Contested Discourses of Ecological Modernization, Green Governmentality and Civic Environmentalism." *Global Environmental Politics* 6, no. 1:50–75.

Baden, John A., ed. 1994. *Environmental Gore: A Constructive Response to* Earth in the Balance. San Francisco: Pacific Research Institute for Public Policy.

Banay, Sophia. 2006. "Hottest Star Hideaways." *Forbes.com,* December 12. Accessed May 26, 2011. http://www.forbes.com/2006/12/12/hottest-star-hideaways-forbeslife-cx_sb_1213hotstarhideaways_slide_7.html?thisSpeed=undefined.

Baram, Marcus. 2007. "An Inconvenient Verdict for Gore: British Court Ruling on Errors in 'An Inconvenient Truth' Resurrects Global Warming Debate." *ABC News,* October 12. Accessed May 25, 2011. http://abcnews.go.com/US/TenWays/story?id=3719791&page=1.

Baudrillard, Jean. 1994. *Simulacra and Simulation.* Trans. Sheila Faria Glaser. Ann Arbor: University of Michigan Press, 1994.

————. 1996. "Disneyworld Company." *Liberation,* March 4. Accessed May 28, 2011. http://www.egs.edu/faculty/jean-baudrillard/articles/disneyworld-company/.

BBC News. 1998. "Disney Animal Deaths Investigated." *BBC News Online,* April 9. Accessed May 27, 2011. http://news.bbc.co.uk/1/hi/world/americas/76154.stm.

Bennett, Tony. 1995. *The Birth of the Museum: History, Theory, Politics.* New York: Routledge.

————. 2005. "Civic Laboratories: Museums, Cultural Objecthood, and the Governance of the Social." *Cultural Studies* 19, no. 5:521–47.

Birkenholtz, Trevor. 2009. "Groundwater Governmentality: Hegemony and Technologies of Resistance in Rajasthan's (India) Groundwater Governance." *Geographical Journal* 175, no. 3:208–20.

Birnbaum Travel Guides, ed. 2005. *Birnbaum's 2006 Walt Disney World: Expert Advice from the Inside Source.* New York: Disney Editions.

Bishop, Matthew. 2007. "The Myth and Magic of Mickey Mouse." *The Economist,* May 30. Accessed May 27, 2011. http://www.economist.com/displaystory.cfm?story_id=8912135#monday.

Bitgood, Stephen C. 1995. "Visitor Circulation: Is There Really a Right-Turn Bias?" *Visitor Behavior* X, no. 1:5.

Blair, Jennifer. 2008. "Media on Ice." *Feminist Media Studies* 8, no. 3:318–23.

Blom, Allard. 2000. "The Monetary Impact of Tourism on Protected Area Management and the Local Economy in Dzanga-Sangha (Central African Republic)." *Journal of Sustainable Tourism* 8, no. 3:175–89.

Booker, Christopher. 2009. "Climate Change: This Is the Worst Scientific Scandal of Our Generation." *Telegraph.co.uk,* November 28. Accessed May 24, 2011. http://www.telegraph.co.uk/comment/columnists/christopherbooker/6679082/Climate-change-this-is-the-worst-scientific-scandal-of-our-generation.html.

Boykoff, Max. 2007. "Flogging a Dead Norm? Newspaper Coverage of Anthropogenic Climate Change in the United States and United Kingdom from 2003 to 2006." *Area* 39, no. 2:1–12.

Boykoff, Max, and Michael K. Goodman. 2009. "Conspicuous Redemption? Reflections on the Promises and Perils of the 'Celebritization' of Climate Change." *Geoforum* 40, no. 3:395–406.

Braun, Bruce. 2002. *The Intemperate Rainforest: Nature, Culture, and Power on Canada's West Coast.* Minneapolis: University of Minnesota Press.

———. 2003. "'On the Raggedy Edge of Risk': Articulations of Race and Nature after Biology." In *Race, Nature, and the Politics of Difference,* ed. Donald Moore, Jake Kosek, and Anand Pandian, 175–203. Durham: Duke University Press.

Breslau, Karen. 2006. "The Resurrection of Al Gore." *Wired.com,* May. Accessed May 27, 2011. http://www.wired.com/wired/archive/14.05/gore.html.

Broder, John M. 2009. "Gore's Dual Role: Advocate and Investor." *New York Times,* November 3. Accessed May 24, 2011. http://www.nytimes.com/2009/11/03/business/energy-environment/03gore.html.

Brooke, J. 1996. "A Montana Town Sees Mine's Gold and Dross." *New York Times,* January 7.

Brower, Matthew. 2005. "Trophy Shots: Early North American Non-human Animal Photography and the Display of Masculine Prowess." *Society and Animals* 13, no. 1:13–31.

Brown, Michael P., with Paul Boyle. 2000. "National Closets: Governmentality, Sexuality, and the Census." In *Closet Space: Geographies of Metaphor from the Body to the Globe,* 88–115. London: Routledge.

Bryman, Alan. 1995. *Disney and His Worlds.* London: Routledge.

Budd, Mike. 2005. "Introduction: Private Disney, Public Disney." In *Rethinking Disney: Private Control, Public Dimensions,* ed. Mike Budd and Max H. Kirsch, 1–33. Middletown: Wesleyan University Press.

Chacko, Sunil, and Henri-Phillippe Sambuc. 2003. "Blockbusters, Traditional Knowledge, and Intellectual Property." *Indigenous Law Bulletin* 5, no. 22. Accessed May 27, 2011. http://kirra.austlii.edu.au/au/journals/ILB/2003/7.html.

Clifford, James. 1995. *Routes: Travel and Translation in the Late Twentieth Century.* Cambridge, Mass.: Harvard University Press.

Cockburn, Alexander, and Jeffrey St. Clair. 2000. *Al Gore: A User's Manual.* London: Verso.

Coleman, Jon T. 2004. *Vicious: Wolves and Men in America.* New Haven, Conn.: Yale University Press.

Conn, Steven. 1998. *Museums and American Intellectual Life, 1876–1926.* Chicago: University of Chicago Press.

Cooke City Chamber of Commerce. n.d. "History of Cooke City." Accessed May 27, 2011. http://www.cookecitychamber.org/cooke_city_montana_history.html.

Cronon, William. 1996. "The Trouble with Wilderness; or, Getting Back to the Wrong Kind of Nature." In *Uncommon Ground: Rethinking the Human Place in Nature,* ed. William Cronon, 69–90. New York: W. W. Norton.

Cruikshank, Julie. 2005. *Do Glaciers Listen? Local Knowledge, Colonial Encounters, and Social Imagination.* Vancouver: University of British Columbia Press.

Daily, Gretchen C., Tore Söderqvist, Sara Aniyar, Kenneth Arrow, Partha Dasgupta, Paul R. Ehrlich, Carl Folke, AnnMari Jansson, Bengt-Owe Jansson, Nils Kautsky, Simon Levin, Jane Lubchenco, Karl-Göran Mäler, David Simpson, David Starrett, David Tilman, and Brian Walker. 2000. "The Value of Nature and the Nature of Value." *Science* 289, no. 5478:395–96.

Dalton, Kathleen. 2002. *Theodore Roosevelt: A Strenuous Life.* New York: Alfred A. Knopf.

Darier, Eric. 1996. "Environmental Governmentality: The Case of Canada's Green Plan." *Environmental Politics* 5, no. 4:585–606.

——— . 1999. "Foucault and the Environment." In *Discourses of the Environment,* ed. Eric Darier, 1–33. Malden, Mass.: Blackwell.

Davey, Gareth, and Peter Henzi. 2004. "Visitor Circulation and Nonhuman Animal Welfare: An Overlooked Variable?" *Journal of Applied Animal Welfare Science* 7, no. 4:243–51.

Davis, Susan G. 1997. *Spectacular Nature: Corporate Culture and the Sea World Experience.* Berkeley: University of California Press.

Dean, Mitchell. 1999. *Governmentality: Power and Rule in Modern Society.* London: Sage.

Delaney, Ted. 1998. "Cold Comes to Yellowstone; But Debate over Fire Policy Still Smolders." *Colorado Springs Gazette Telegraph,* November 13, A1.

Desmond, Jane C. 1999. *Staging Tourism: Bodies on Display from Waikiki to Sea World.* Chicago: University of Chicago Press.

Disney Wildlife Conservation Fund. n.d. "Instructions to Apply for a Disney Wildlife Conservation Fund Award." Accessed May 22, 2011. http://disney .go.com/disneyhand/environmentality/dwcf/apply.html.

Disney Worldwide Outreach. n.d. "Environmentality: Disney's Wildlife Conservation Fund—Who Supports the Fund?" Accessed May 22, 2011. http://disney .go.com/disneyhand/environmentality/dwcf/who.html.

Duffy, Rosaleen. 2002. *A Trip Too Far: Ecotourism, Politics and Exploitation.* London: Earthscan.

Dunaway, Finis. 2000. "Hunting with the Camera: Nature Photography, Manliness, and Modern Memory, 1890–1930." *Journal of American Studies* 34:207–30.

Eldredge, Niles. 1998. *Life in the Balance: Humanity and the Biodiversity Crisis.* Princeton: Princeton University Press.

Emel, Jody. 2002. "An Inquiry into the Green Disciplining of Capital." *Environment and Planning A* 34, no. 5:827–43.

Energy Star. n.d. "Light Bulbs (CFLs) for Consumers." Accessed May 25, 2011. http://www.energystar.gov/index.cfm?fuseaction=find_a_product.show ProductGroup&pgw_code=LB.

Erickson, Bruce. 2003. "The Colonial Climbs of Mount Trudeau: Thinking Masculinity through the Homosocial." *TOPIA* 9:67–82.

Escobar, Arturo. 1996. "Constructing Nature: Elements for a Poststructural Political Ecology." In *Liberation Ecologies: Environment, Development, Social Movements,* ed. Richard Peet and Michael Watts, 48–68. New York: Routledge.

Faery, Rebecca B. 1999. *Cartographies of Desire: Captivity, Race, and Sex in the Shaping of an American Nation.* Norman: University of Oklahoma Press.

Fairclough, Norman. 2003. *Analysing Discourse: Textual Analysis for Social Research.* London: Routledge.

Falk, J. H., and L. D. Dierking. 1992. "Global Warming: Understanding the Forecast (Exhibit Review)." *Museum Anthropology* 16, no. 3:72–78.

Farley, Rebecca. 2005. "'By Endurance We Conquer': Ernest Shackleton and the Performance of White Male Hegemony." *International Journal of Cultural Studies* 8, no. 2:231–54.

Fennell, David A. 1999. *Ecotourism: An Introduction.* London: Routledge.

Fishman, B. 2010. "Dispatch: Cooke City, Montana." *TMagazine blogs,* June 28. Accessed May 27, 2011. http://tmagazine.blogs.nytimes.com/2010/06/28/dispatch-cooke-city-montana/.

Fjellman, Stephen M. 1992. *Vinyl Leaves: Walt Disney World and America.* Boulder, Colo.: Westview Press.

Flocker, Michael. 2003. *The Metrosexual Guide to Style: A Handbook for the Modern Man.* Cambridge, Mass.: Da Capo Press.

Fogelsong, Richard E. 2001. *Married to the Mouse: Walt Disney World and Orlando.* New Haven: Yale University Press.

Foucault, Michel. 1980. "The Eye of Power." In *Power/Knowledge: Selected Interviews and Other Writings, 1972–1977,* ed. Colin Gordon, 146–56. New York: Pantheon.

———. 1982. "Afterword: The Subject and Power." In *Michel Foucault: Beyond Structuralism and Hermeneutics,* ed. Hubert L. Dreyfus and Paul Rabinow, 208–26. New York: Harvester-Wheatsheaf.

———. 1984. "On the Genealogy of Ethics: An Overview of Work in Progress." In *The Foucault Reader,* ed. Paul Rabinow, 340–72. New York: Pantheon Books.

———. 1990. *The History of Sexuality: Volume 1.* New York: Vintage.

———. 1991. "Governmentality." In *The Foucault Effect: Studies in Governmentality,* ed. Graham Burchell, Colin Gordon, and Peter Miller, 87–104. Chicago: University of Chicago Press.

———. 1994. "The Political Technology of Individuals." In *Power (The Essential Works of Foucault, 1954–1984,* vol. 3), ed. James D. Faubion, 403–17. New York: The New Press.

———. 1995. *Discipline and Punish: The Birth of the Prison.* New York: Vintage.

————. 1997. "Technologies of the Self." In *Ethics: Subjectivity and Truth* (*The Essential Works of Foucault, 1954–1984*, vol. 1), ed. Paul Rabinow, 223–51. New York: The New Press.

————. 1998. "Different Spaces." In *Aesthetics, Method, and Epistemology* (*The Essential Works of Foucault, 1954–1984*, vol. 2), ed. James D. Faubion, 175–85. New York: The New Press.

————. 2001. *Fearless Speech*, recorded, compiled, and edited by Joseph Pearson. New York: Semiotext(e).

————. 2003a. *Society Must Be Defended: Lectures at the College de France, 1975–76*. New York: Picador.

————. 2003b. *The Birth of the Clinic.* London: Routledge.

————. 2004. *The Order of Things: An Archaeology of the Human Sciences.* London: Routledge.

————. 2007. *Security, Territory, Population: Lectures at the College de France, 1977–78.* Hampshire: Palgrave MacMillan.

————. 2010. *The Birth of Biopolitics: Lectures at the College de France, 1978–79.* New York: Picador.

Frankenberg, Ruth. 1997. *Displacing Whiteness: Essays in Social and Cultural Criticism.* Durham: Duke University Press.

GAIAM. 2005. *GAIAM Annual Report 2005.* Accessed November 1, 2007. http://corporate.gaiam.com/crp_Earnings.asp?Type=investor.

Gallup. 2009. "Increased Number Think Global Warming is 'Exaggerated.'" March 11. Accessed May 27, 2011. http://www.gallup.com/poll/116590/increased-number-think-global-warming-exaggerated.aspx.

Germic, Stephen A. 2001. *American Green: Class, Crisis, and the Deployment of Nature in Central Park, Yosemite, and Yellowstone.* Lanham, Md.: Lexington Books, 2001.

Gilbert, Helen. 2004. "Ecotourism: A Colonial Legacy?" Accessed May 24, 2011. http://espace.library.uq.edu.au/eserv.php?pid=UQ:10761&dsID=hg_eco tourism.pdf.

Giroux, Henri. A. 1993. "Beyond the Politics of Innocence: Memory and Pedagogy in the 'Wonderful World of Disney.'" *Socialist Review* 23, no. 2:79–107.

Glaser, Jane. R., and Artemis A. Zenetou. 1994. *Gender Perspectives: Essays on Women in Museums.* Washington, D.C.: Smithsonian Institution Press.

Glover, Dominic. 2010. "The Corporate Shaping of GM Crops as Technology for the Poor." *Journal of Peasant Studies* 37, no. 1:67–90.

Gore, Albert. 1992. *Earth in the Balance: Ecology and the Human Spirit.* New York: Rodale.

————. 2006. "The Moment of Truth." *Vanity Fair,* May. Accessed May 25, 2011. http://www.vanityfair.com/politics/features/2006/05/gore200605?print able=true.

————. 2009. *Our Choice: A Plan to Solve the Climate Crisis.* Emmaus, Pa.: Rodale.

————. 2010. "We Can't Wish Away Climate Change." *New York Times,* February 28, 11.

Gottlieb, Robert S. 2006. *A Greener Faith: Religious Environmentalism and Our Planet's Future.* Oxford: Oxford University Press.

Gramsci, Antonio. 1992. *Prison Notebooks.* New York: Columbia University Press.

Grant Thornton LLP. 2009. "The American Museum of Natural History Consolidated Financial Statements for the Years Ended June 30, 2009 and 2008." AMNH. Accessed May 27, 2011. http://www.amnh.org/about/pdf/amnh_6_30_09_08.pdf.

Griffin, Susan. 2005. "Curiouser and Curiouser: Gay Days at the Disney Theme Parks." In *Rethinking Disney: Private Control, Public Dimensions,* ed. Mike Budd and Max H. Kirsch, 125–50. Middletown: Wesleyan University Press.

Haggerty, Kevin D., and Richard V. Ericson. 2000. "The Surveillant Assemblage." *British Journal of Sociology* 15, no. 4:605–22.

Hannah, Matthew. 2000. *Governmentality and the Mastery of Territory in Nineteenth-Century America.* Cambridge: Cambridge University Press.

Haraway, Donna J. 1989. "Teddy Bear Patriarchy: Taxidermy and the Garden of Eden, New York City, 1908–36." In *Primate Visions: Gender, Race, and Nature in the World of Modern Science,* 26–58. New York: Routledge.

————. 1991. *Simians, Cyborgs, and Women: The Reinvention of Nature.* London: Free Association Books.

————. 1992. "The Promise of Monsters: A Regenerative Politics for Inappropriate/d Others." In *Cultural Studies,* ed. Lawrence Grossberg, Cary Nelson, and Paula Treichler, 275–332. New York: Routledge.

————. 1996. "Universal Donors in a Vampire Culture: It's All in the Family: Biological Kinship Categories in the Twentieth-Century United States." In *Uncommon Ground: Rethinking the Human Place in Nature,* ed. William Cronon, 321–66. New York: W. W. Norton and Co.

————. 2003. "For the Love of a Good Dog: Webs of Action in the World of Dog Genetics." In *Race, Nature, and the Politics of Difference,* ed. Donald S. Moore, Jake Kosek, and Anand Pandian, 254–95. Durham: Duke University Press.

————. 2007. *When Species Meet.* Minneapolis: University of Minnesota Press.

Hardin, Rebecca. 2000. "Translating the Forest: Tourism, Trophy Hunting, and the Transformation of Forest Use in Southwestern Central African Republic." PhD diss., Yale University.

Harmel, Karen. 2006. "Walt Disney World: The Government's Tomorrowland?" *News21,* September 1. Accessed May 27, 2011. http://newsinitiative.org/story/2006/09/01/walt_disney_world_the_governments.

Hermanson, Scott. 2005. "Truer Than Life: Disney's Animal Kingdom." In *Rethinking Disney: Private Control, Public Dimensions*, ed. Mike Budd and Max H. Kirsch, 199–227. Middletown: Wesleyan University Press.

Hetherington, Kevin. 1997. "Museum Typology and the Will to Connect." *Journal of Material Culture* 2, no. 2:199–220.

Honey, Martha. 1999. *Ecotourism and Sustainable Development: Who Owns Paradise?* Washington, D.C.: Island Press.

Hooper-Greenhill, Eileen. 2001. *Museums and the Shaping of Knowledge*. London: Routledge.

An Inconvenient Truth. 2006. DVD. Directed by David Guggenheim with Al Gore. Paramount Pictures.

The International Ecotourism Society. 2000. "Ecotourism Statistical Fact Sheet: 2000." Accessed August 22, 2007. http://www.ecotourism.org/WebModules/ WebMember/MemberApplication/onlineLib/MemberApplication/online Lib/Uploaded/Ecotourism%20Factsheet%202000.pdf.

———. 2006. "TIES Global Ecotourism Fact Sheet." Accessed May 24, 2011. http://www.ecotourism.org/atf/cf/%7B82a87c8d-0b56-4149-8b0a-c4aaced1cd38%7D/TIES%20GLOBAL%20ECOTOURISM%20FACT%20 SHEET.PDF.

Johnson, Laura. 2009. "(Environmental) Rhetorics of Tempered Apocalypticism in *An Inconvenient Truth*." *Rhetoric Review* 28, no. 1: 29–46.

Jones, David, Andrew Watkins, Karl Braganza, and Michael Coughlan. 2007. "*The Great Global Warming Swindle*: A Critique." *Bulletin of the Australian Meteorological and Oceanographic Society* 20. Accessed May 25, 2011. http://www .csiro.au/files/files/pfb4.pdf.

Jones, Karen R. 2002. *Wolf Mountains: A History of Wolves Along the Great Divide*. Calgary: University of Calgary Press.

Kadanoff, Leo P. 2002. "Sue's Several Heads: The Evolution of the Natural History Museum." *Perspectives in Biology and Medicine* 45, no. 2:272–80.

Katz, Cindi. 1998. "Whose Nature, Whose Culture? Private Productions of Space and the Preservation of Nature." In *Remaking Reality: Nature at the End of the Millennium*, ed. Bruce Braun and Noel Castree, 46–63. New York: Routledge.

Kelemen, Peter. 2009. "What East Anglia's E-mails Really Tell Us about Climate Change." *Popular Mechanics*, December 18. Accessed May 24, 2011. http:// www.popularmechanics.com/science/earth/4338343.html?page=2.

King, Margaret J. 1991. "The Theme Park Experience: What Museums Can Learn from Mickey Mouse." *The Futurist* 25, no. 6:24–31.

———. 1996. "The Audience in the Wilderness: The Disney Nature Films." *Journal of Popular Film and Television* 24: 60–68.

King, Thomas. 2003. *The Truth about Stories: A Native Narrative*. Toronto: House of Anansi Press.

King, Ynestra. 1997. "Managerial Environmentalism, Population Control and the New National Insecurity: Towards a Feminist Critique." *Political Environments: A Publication of Committee on Women, Population, & The Environment* 5 (Fall). Accessed May 28, 2011. http://www.cwpe.org/node/135.

Kobayashi, Audrey, and Linda Peake. 2000. "Racism Out of Place: Thoughts on Whiteness and an Anti-racist Geography in the New Millennium." *Annals of the Association of American Geographers* 90, no. 2:392–403.

Kuhn, Thomas S. 1996. *The Structure of Scientific Revolutions.* 3rd ed. Chicago: University of Chicago Press.

Larry King Live. 2009. November 13. Al Gore on Larry King Live. Atlanta: CNN. Accessed on May 26, 2011. http://blog.algore.com/2009/11/the_tonight_show_and_larry_kin.html.

Latour, Bruno. 1993. *We Have Never Been Modern.* Cambridge, Mass.: Harvard University Press.

Law, John. 1999. "After ANT: Complexity, Naming and Typology." In *Actor Network Theory and After,* ed. John Law and John Hassard, 1–14. Oxford: Blackwell Publishing.

Lees, Loretta. 2004. "Urban Geography: Discourse Analysis and Urban Research." *Progress in Human Geography* 28, no. 1:101–7.

Legates, David R. 2007. "*An Inconvenient Truth:* A Focus on Its Portrayal of the Hydrological Cycle." *GeoJournal* 70:15–19.

Leiserowitz, Anthony, Edward Maibach, and Connie Roser-Renouf. 2010. "Climate Change in the American Mind: Americans' Global Warming Beliefs and Attitudes in January 2010." Yale University and George Mason University. New Haven: Yale Project on Climate Change. Accessed November 6, 2010. http://environment.yale.edu/uploads/AmericanGlobalWarmingBeliefs2010.pdf.

Leonard, Tom. 2008. "Al Gore's Electricity Bill Goes through the (Insulated) Roof." *The Telegraph,* June 18. Accessed July 6, 2010. http://www.telegraph.co.uk/news/worldnews/northamerica/usa/2153179/Al-Gores-electricity-bill-goes-through-the-insulated-roof.html.

Lequm, Judd. 2006. "Exxon-Backed Pundit Compares Gore to Nazi Propagandist." Thinkprogress.org, May 23. Accessed March 16, 2010. http://thinkprogress.org/2006/05/23/gore-movie-g/.

Lopez, Barry H. 1978. *Of Wolves and Men.* New York: Charles Scribner's Sons.

Luke, Timothy W. 1997. "The (Un)wise (Ab)use of Nature: Environmentalism as Globalized Consumerism." Paper presented at the annual meeting of the International Studies Association, Toronto, Ontario, March 18–22.

———. 1999. "Environmentality as Green Governmentality." In *Discourses of the Environment,* ed. Eric Darier, 121–51. Malden, Mass.: Blackwell.

———. 2002a. "Museum Pieces: Politics and Knowledge at the American Museum of Natural History." In *Museum Politics: Power Plays at the Exhibition*, 100–123. Minneapolis: University of Minnesota Press.

———. 2002b. "The Missouri Botanical Garden: Sharing Knowledge about Plants to Preserve and Enrich Life." In *Museum Politics: Power Plays at the Exhibition*, 124–45. Minneapolis: University of Minnesota Press.

———. 2003. "On the Political Economy of Clayoquot Sound: The Uneasy Transition from Extractive to Attractive Models of Development." In *A Political Space: Reading the Global through Clayoquot Sound*, ed. Warren Magnusson and Karena Shaw, 91–112. Minneapolis: University of Minnesota Press.

———. 2008. "The Politics of True Inconvenience or Inconvenient Truth: Struggles Over How to Sustain Capitalism, Democracy, and Ecology in the 21st Century." *Environment and Planning A* 40, no. 8:1811–24.

Lyman, Francesca. 1998. *Inside the Dzanga-Sangha Rain Forest*. New York: Workman Publishing.

MacCannell, Dean. 1989. *The Tourist: A New Theory of the Leisure Class*. New York: Schocken Books.

Macdonald, Sharon. 1995. "Editorial: Science on Display." *Science as Culture* 5, no. 1:7–11.

———, ed. 2000. *The Politics of Display: Museums, Science, Culture*. New York: Routledge.

———. 2002. *Behind the Scenes at the Science Museum*. Oxford: Berg.

Macdonald, Sharon, and Gordon Fyfe, eds. 1996. *Theorizing Museums: Representing Identity and Diversity in a Changing World*. Oxford: Blackwell.

Mackay, David. 1996. "Agents of Empire: The Banksian Collectors and Evaluation of New Lands." In *Visions of Empire: Voyages, Botany, and Representations of Nature*, ed. David P. Miller and Peter H. Reill, 38–57. Cambridge: Cambridge University Press.

Macnaghten, Phil. 2003. "Embodying the Environment in Everyday Life Practices." *The Sociological Review* 51, no. 1:62–84.

Magoc, Chris J. 1999. *Yellowstone: The Creation and Selling of an American Landscape, 1870–1903*. Albuquerque: University of New Mexico Press.

Malmberg, Melody. 1998. *The Making of Disney's Animal Kingdom™ Theme Park*. New York: Hyperion.

McClintock, Anne. 1995. *Imperial Leather: Race, Gender, and Sexuality in the Colonial Contest*. New York: Routledge.

McClure, Robert, and Lisa Stiffler. 2007. "Federal Way Schools Restrict Gore Film: 'Inconvenient Truth' Called Too Controversial." *Seattle Post-Intelligencer*, January 11. Accessed June 29, 2010. http://www.seattlepi.com/local/299253_inconvenient11.html.

McHoul, Alec, and Wendy Grace. 1993. *A Foucault Primer: Discourse, Power, and the Subject*. New York: New York University Press.

Media Matters for America. 2006. "Beck on *An Inconvenient Truth*: 'It's like Hitler.'" June 8. Accessed May 25, 2011. http://mediamatters.org/mmtv/200606080005.

Miller, Toby. 1993. *The Well-Tempered Self: Citizenship, Culture, and the Postmodern Subject.* Baltimore: Johns Hopkins University Press.

Mills, Sara. 2003. *Michel Foucault.* London: Routledge.

Mitchell, Heidi. S. 2006. "Easy Being Green." *T Magazine,* September 24, 14. Accessed May 28, 2011. http://travel2.nytimes.com/2006/09/24/travel/tmagazine/24green.html.

Mitchell, Timothy. 1988. *Colonising Egypt.* New York: Cambridge University Press.

———. 2002. *The Rule of Experts: Egypt, Techno-politics, Modernity.* Berkeley: University of California Press.

Mittman, Gregg. 2003. *Reel Nature: America's Romance with Wildlife on Film.* Cambridge, Mass.: Harvard University Press.

Monbiot, George. 2007. "Don't Let the Truth Stand in the Way of a Red-Hot Debunking of Climate Change." *Guardian,* March 13. Accessed June 24, 2010. http://www.guardian.co.uk/commentisfree/2007/mar/13/science.media.

Moore, Donald S., Jake Kosek, and Anand Pandian, eds. 2003. *Race, Nature, and the Politics of Difference.* Durham: Duke University Press.

Moran, Katy. 2000. "Bioprospecting: Lessons from Benefit-Sharing Experiences." *International Journal of Biotechnology* 2, nos. 1–3:132–44.

Morrissey, Brian. 1998. "The Making of a Rainforest." *Natural History* 107, no. 5:56–62.

Murray, Robin, and Joseph Heumann. 2007. "Al Gore's 'An Inconvenient Truth' and Its Skeptics: A Case of Environmental Nostalgia." *JumpCut* 49. Accessed June 24, 2010. http://www.ejumpcut.org/archive/jc49.2007/inconvenTruth/index.html.

Natural Habitat Adventures. 2007. *Natural Habitat Adventures 2007–2008 Catalogue of the World's Greatest Nature Expeditions.* Boulder, Colo.: Natural Habitat Adventures.

———. n.d.1. "Destinations." Accessed February 11, 2008. http://www.nathab.com/destinations/index.aspx.

———. n.d.2. "The Habitat Club—New and Improved." Accessed February 11, 2008. http://www.nathab.com/habitat-club.

———. n.d.3. "Hidden Yellowstone and Grand Teton." Accessed February 11, 2008. http://www.nathab.com/america/hidden-yellowstone.

———. n.d.4. "Natural Habitat Adventures Is Proud to Be the World's First 100% Carbon Neutral Travel Company!" Accessed May 27, 2011. http://www.nathab.com/carbon-offsetting/index.aspx?pageID=108.

Nichols, Bill. 1991. *Representing Reality: Issues and Concepts in Documentary.* Bloomington: Indiana University Press.

————. 2001. *Introduction to Documentary.* Bloomington: Indiana University Press.

Nielsen Company and Environmental Change Institute (Oxford University). 2007a. "Climate Change and Influential Spokespeople: A Global Nielsen Online Survey." Accessed June 25, 2010. http://www.eci.ox.ac.uk/publications/down loads/070709nielsen-celeb-report.pdf.

————. 2007b. "Global Consumers Vote Al Gore, Oprah Winfrey and Kofi Annan Most Influential to Champion Global Warming Cause: Nielsen Survey." Accessed May 25, 2011. http://nz.acnielsen.com/news/GlobalWarm ing_Ju107.shtml.

Nielsen-Gammon, John W. 2007. "*An Inconvenient Truth:* The Scientific Argument." *GeoJournal* 70:21–26.

Nordhaus, Ted, and Michael Shellenberger. 2007. *Break Through: From the Death of Environmentalism to the Politics of Possibility.* Boston: Houghton Mifflin.

Noss, Andrew J. 1997. "Challenges to Conservation with Community Development in Central African Forests." *Oryx* 31, no. 3:180–88.

NPS (National Park Service). 2009a. "Grand Teton National Park Public Visitation Report." Accessed August 25, 2010. http://www2.nature.nps.gov/stats/.

————. 2009b. "Yellowstone National Park Public Visitation Report." Accessed August 25, 2010. http://www2.nature.nps.gov/stats/.

O'Brien, Tim. 1998. "Disney's Animal Kingdom Hit Capacity Opening Day." *Amusement Business* 110, no. 18:1–2.

Ogden, Jackie, Kathy Lehnhardt, Jill Mellen, Lynn Dierking, Leslie Adelman, Kyle Burks, and Lance Miller. 2001. "Conservation Station and Beyond: Experiences at Disney's Animal Kingdom That Make a Difference." *Informal Learning Review* 51:18–19.

O'Malley, Pat, Lorna Weir, and Clifford Shearing. 1997. "Governmentality, Criticism, Politics." *Economy and Society* 26, no. 4:501–17.

Omi, Michael, and Howard Winant. 1994. *Racial Formation in the United States: From the 1960s to the 1990s.* 2nd ed. New York: Routledge.

Pearce, Susan M. 1992. "Problems of Power." In *Museums, Objects, and Collections: A Cultural Study,* 228–55. Leicester: Leicester University Press.

Pew Research Center for the People and the Press. 2006. "Little Consensus on Global Warming: Partisanship Drives Opinion." July 12. Accessed March 15, 2010. http://people-press.org/report/280/little-consensus-on-global-warming.

Pluskoski, Aleksander. 2006. *Wolves and Wilderness in the Middle Ages.* Rochester, N.Y.: Boydell.

Potter, Emily, and Candice Oster. 2008. "Communicating Climate Change: Public Responsiveness and Matters of Concern." *Media International Australia* 127:116–26.

Powhatan Renape Nation. n.d. "The Pocahontas Myth." Accessed February 11, 2008. http://www.powhatan.org/pocc.html.

Poynton, Cate, and Allison Lee. 2000. "Culture and Text: An Introduction." In *Culture and Text: Discourse and Methodology in Social Research and Cultural Studies,* ed. Cate Poynton and Allison Lee, 1–18. Lanham, Md.: Rowman and Littlefield.

Pratt, Mary Louise. 1992. *Imperial Eyes: Travel Writing and Transculturation.* New York: Routledge.

Price, Jenny. 1996. "Looking for Nature at the Mall: A Field Guide to the Nature Company." In *Uncommon Ground: Rethinking the Human Place in Nature,* ed. William Cronon, 186–203. New York: W. W. Norton and Co.

The Project on Disney. 1995. *Inside the Mouse: Work and Play at Disney World.* Durham: Duke University Press.

Prudham, Scott. 2007. "The Fictions of Autonomous Invention: Accumulation by Dispossession, Commodification and Life Patents in Canada." *Antipode* 39, no. 3:406–29.

Putnam, Robert D. 1993. *Making Democracy Work: Civic Traditions in Modern Italy.* Princeton: Princeton University Press.

Quinn, Stephen C. 2006. *Windows on Nature: The Great Habitat Dioramas of the American Museum of Natural History.* New York: Harry N. Abrams (in association with the American Museum of Natural History).

Rabinow, Paul. 1984. Introduction to *The Foucault Reader,* ed. Paul Rabinow, 3–29. New York: Pantheon Books.

Rabinow, Paul, and Nikolas Rose. 2003. "Thoughts on the Concept of Biopower Today." Last revised December 10. Accessed May 27, 2007. http://caosmosis .acracia.net/wp-content/uploads/2009/04/rabinow-y-rose-biopowertoday -1.pdf.

Remis, Melissa J., and Rebecca Hardin. 2009. "Transvalued Species in an African Forest." *Conservation Biology* 23, no. 6:1588–96.

Renov, Michael, ed. 1993. *Theorizing Documentary.* New York: Routledge.

Renshaw, Sal. 2010. "Review of Chloë Taylor, The Culture of Confession from Augustine to Foucault: A Genealogy of the 'Confessing Animal.'" *Foucault Studies* 8:174–79.

Rexer, Lyle, and Rachel Klein. 1995. *American Museum of Natural History: 125 Years of Expedition and Discovery.* New York: Harry N. Abrams (in association with the American Museum of Natural History).

Rosaldo, Renato. 1989. *Culture and Truth: The Remaking of Social Analysis.* Boston: Beacon Press.

Rose, Nikolas. 1996. *Inventing Our Selves: Psychology, Power, and Personhood.* New York: Cambridge University Press.

———. 2006. *The Politics of Life Itself: Biomedicine, Power, and Subjectivity in the Twenty-First Century.* Princeton: Princeton University Press.

Rosteck, Thomas, and Thomas S. Frentz. 2009. "Myth and Multiple Readings in

Environmental Rhetoric: The Case of *An Inconvenient Truth.*" *Quarterly Journal of Speech* 95, no. 1:1–19.

Rutherford, Paul. 1994. "Policing Nature: Ecology, Natural Sciences, and Biopolitics." *Foucault: The Legacy Conference.* Surfer's Paradise, Australia, 546–62.

———. 1999. "The Entry of Life into History." In *Discourses of the Environment,* ed. Eric Darier, 37–62. Malden, Mass.: Blackwell.

Rutherford, Stephanie. 2007. "Green Governmentality: Insights and Opportunities in the Study of Nature's Rule." *Progress in Human Geography* 31, no. 3:291–307.

Sachs, Wolfgang. 1999. *Planet Dialectics: Explorations in Environment and Development.* London: Zed Books.

Sagi, Douglas. 1992. "Scientists Demolish Lemming Legends." *Vancouver Sun,* February 21, D2.

Said, Edward. 1979. *Orientalism.* New York: Vintage Books.

Sandilands, Catriona. 2003. "Between the Local and the Global: Clayoquot Sound and Simulacral Politics." In *A Political Space: Reading the Global through Clayoquot Sound,* 139–67. Minneapolis: University of Minnesota Press.

Schabecoff, Philip. 1988. "Park and Forest Service Chiefs Assailed on Fire Policy." *New York Times,* September 10, 1.

Schaffer, Marguerite S. 2001. *See America First: Tourism and National Identity, 1880–1940.* Washington, D.C.: Smithsonian Institution Press.

Schiebinger, Londa. 2004. *Plants and Empire: Colonial Bioprospecting in the Atlantic World.* Cambridge, Mass.: Harvard University Press.

Scott, David. 1999. *Refashioning Futures: Criticism after Postcoloniality.* Princeton: Princeton University Press.

Sheets-Pyenson, Susan. 1988. *Cathedrals of Science: The Development of Colonial Natural History Museums during the Late Nineteenth Century.* Kingston: McGill-Queen's University Press.

Sherman, Daniel J., and Irit Rogoff. 1994. "Introduction: Frameworks for Critical Analysis." In *Museum Culture: Histories, Discourses, Spectacles,* ed. Daniel J. Sherman and Irit Rogoff, ix–xx. Minneapolis: University of Minnesota Press.

Simpson, Mark. 2002. "Meet the Metrosexual." *Salon.com,* July 22. Accessed September 11, 2010. http://marksimpson.com/pages/journalism/metrosexual_beckham.html.

———. 2006. "Here Come the Mirror Men." Mark Simpson Web site. Accessed September 11, 2010. http://marksimpson.com/pages/journalism/mirror_men .html.

Skaggs, Jackie. 2000. "Creation of Grand Teton National Park (A Thumbnail History)." January. Accessed August 22, 2007. http://www.nps.gov/grte/ planyourvisit/upload/creation.pdf.

Smart, Barry. 1985. *Michel Foucault.* Chichester: Ellis Harwood Ltd.

Smith, Douglas W., Daniel R. Stahler, Debra S. Guernsey, with Matthew Metz, Abigail Nelson, Erin Albers, and Richard McIntyre. 2007. *Yellowstone Wolf Project: Annual Report, 2006* (No. YCR-2007-01.). Yellowstone National Park, Wyo.: National Park Service, Yellowstone Center for Resources.

Society for Environmental Graphic Design. 1999. "1999 Design Winners: Hall of Biodiversity, American Museum of Natural History." Accessed August 13, 2006. http://www.segd.org/awards/1999.html.

Sontag, Susan. 1978. *On Photography.* New York: Farrar.

Sorkin, Michael. 1992. "See You in Disneyland." In *Variations on a Theme Park: The New American City and the End of Public Space,* 205–32. New York: Hill and Wang.

Spence, Mark D. 1999. *Dispossessing the Wilderness: Indian Removal and the Making of the National Parks.* Oxford: Oxford University Press.

Spencer, Roy. 2007. "*An Inconvenient Truth*: Blurring the Lines between Science and Science Fiction." *GeoJournal* 70, no. 1:11–14.

Stegner, Wallace. 1983. "The Best Idea We Ever Had." *Wilderness* 46:4–13.

Stine, Jeffrey K. 2002. "Placing Environmental History on Display." *Environmental History* 7, no. 4:566–88.

Strathern, Marilyn. 2004. *Partial Connections.* Walnut Creek, Calif.: AltaMira Press.

TEA and AECOM (Themed Entertainment Association and AECOM Economics). 2009. "2009 Theme Index: The Global Attractions Attendance Report." Accessed August 23, 2010. http://www.aecom.com/deployedfiles/Internet/Capabilities/2009%20Theme%20Index%20Final%20004271o_for%2oscreen .pdf.

Teslow, Tracy L. 1995. "Representing Race: Artistic and Scientific Realism." *Science as Culture* 5, no. 1:12–38.

Threadgold, Terry. 2000. "Poststructuralism and Discourse Analysis." In *Culture and Text: Discourse and Methodology in Social Research and Cultural Studies,* ed. Allison Lee and Cate Poynton, 40–58. Lanham, Md.: Rowman and Littlefield.

Tierney, John. 1998. "Going Where a Lot of Other Dudes with Really Great Equipment Have Gone Before." *New York Times Magazine,* July 26, 18–23, 33–34, 46–48.

Tilton, Robert S. 1994. *Pocahontas: The Evolution of an American Narrative.* Cambridge: Cambridge University Press.

Trinh Minh-ha. 1993. "The Totalizing Quest of Meaning." In *Theorizing Documentary,* ed. M. Renov, 90–107. New York: Routledge.

———. 1998. "Inappropriate/d Artificiality: An Interview with Marina Grzinic." Accessed February 18, 2008. http://www.trinhminh-ha.com/.

Turner, Frederick J. 1986. *The Significance of the Frontier in American History.* Tuscon: University of Arizona Press. (Orig. pub. 1893.)

Turque, Bill. 2000. *Inventing Al Gore: A Biography.* Boston: Houghton Mifflin Co.

UNWTO (United Nations World Tourism Organization). 2007. "Tourism High-lights, 2007 Edition." Accessed August 29, 2007. http://unwto.org/facts/eng/pdf/highlights/highlights_07_eng_hr.pdf.

Urry, John. 1992. "The Tourist Gaze and the 'Environment.'" *Theory, Culture, and Society* 9, no. 3:1–26.

————. 2002. *The Tourist Gaze*. 2nd ed. Thousand Oaks, Calif.: Sage Publications.

U.S. Bureau of Labor Statistics. 2009. "State Occupational Employment and Wage Estimates." Bureau of Labor Statistics, May. Accessed May 27, 2011. http://www.bls.gov/oes/current/oes_mt.htm#41-0000.

U.S. Census Bureau. 2006. "Fact Sheet: Cooke City, Montana." Accessed October 15, 2007. http://factfinder.census.gov/servlet/SAFFFacts?_event=&geo_id=16000US3017359&_geoContext=01000US%7C04000US30%7C16000US3017359&_street=&_county=cooke+city&_cityTown=cooke+city&_state=04000US30&_zip=&_lang=en&_sse=on&ActiveGeoDiv=&_useEV=&pctxt=fph&pgsl=160&_submenuId=factsheet_1&ds_name=null&_ci_nbr=null&qr_name=null&reg=null%3Anull&_keyword=&_industry=&show_2003_tab=&redirect=Y.

Waitt, Gordon, and Lauren Cook. 2007. "Leaving Nothing but Ripples on the Water: Performing Ecotourism Natures." *Social and Cultural Geography* 8, no. 4:535–50.

Wali, Alaka. 1999. "Anthropology as the Missing Link: Representing and Advocating for Nature in Natural History Museums." *Culture and Agriculture* 21, no. 3:44–52.

Wallace, Joseph E. 2000. *A Gathering of Wonders: Behind the Scenes at the American Museum of Natural History*. New York: St. Martin's Press.

Walt Disney Company. 2005a. "Disney's Commitment to Conservation." Lake Buena Vista, Fla.: Walt Disney World Conservation Initiatives.

————. 2005b. "Disney's Enviroport: Annual Environmental Report for the Walt Disney Company." Accessed February 8, 2007. http://amedia.disney.go.com/environmentality/enviroports/twdc_2005_enviroreport_a.pdf.

————. 2006. "The Walt Disney Company 2006 Annual Report." Accessed February 8, 2007. http://corporate.disney.go.com/investors/annual_reports/2006/index.html.

————. 2009. "The Walt Disney Company Fiscal Year 2009 Annual Financial Report and Shareholder Letter." Accessed May 22, 2011. http://amedia.disney.go.com/investorrelations/annual_reports/WDC-10kwrap-2009.pdf.

Walt Disney World Public Affairs. n.d.. "Disney's Environmentality." Accessed May 25, 2011. http://www.wdwpublicaffairs.com/LinksTemplateCal.aspx?PageId=5f0d0e9e-3f31-444b-b65b-a7010c369fcd.

Wasko, Janet. 2001. "Is It a Small World, After All?" In *Dazzled by Disney? The*

Global Disney Audiences Project, ed. Janet Wasko, Mark Phillips, and Eileen. R. Meehan, 3–28. London: Leicester University Press.

WDW Magic. 2006. "Expedition Everest Opening Day Press Release." April 7. Accessed May 23, 2007. http://www.wdwmagic.com/Attractions/Expedition -Everest/News/07Apr2006-Expedition-Everest-Opening-Day-press-release .htm.

WDW News—Walt Disney World Resort. 2005a. "Conservation International and Discovery Networks Join Disney on Scientific and Cultural Journey into the Realm of the Yeti." Accessed May 23, 2007. http://www.wdwnews.com/ ViewPressRelease.aspx?PressReleaseID=100334.

———. 2005b. "Disney's Animal Kingdom Fact Sheet." Accessed May 5, 2006. http://www.wdwnews.com/ViewPressRelease.aspx?PressReleaseID=99795.

———. 2005c. "Disney's Animal Kingdom Fun Facts." Accessed May 5, 2006. http://www.wdwnews.com/ViewPressRelease.aspx?PressReleaseID= 100049.

———. 2005d. "Tree of Life Stories in Sculpture at Disney's Animal Kingdom." Accessed May 5, 2006. http://www.wdwnews.com/ViewPressRelease.aspx? PressReleaseID=99800.

———. 2006a. "Animal Programs Team Works to Protect, Propagate and Preserve Animals and Habitats." Accessed May 6, 2006. http://www.wdwnews .com/ViewPressRelease.aspx?PressReleaseID=100050.

———. 2006b. "Conservation Celebrities Recognized as Disney Fund Reaches Milestone $10 Million Investment in Wildlife." Accessed May 6, 2006. http:// www.wdwnews.com/ViewPressRelease.aspx?PressReleaseID=103294.

———. 2006c. "Conservation International and Disney Discover New Species in the 'Realm of the Yeti.'" Accessed May 6, 2006. http://www.wdwpublic affairs.com/ContentDrillDownPrintable.aspx?DisplayItem=1b62eb27- 1b38-44da-9d46-38d7a3095d6e.

———. 2006d. "Respected Authorities Share Their Experiences with Disney." Accessed February 14, 2007. http://www.wdwnews.com/ViewPressRelease .aspx?PressReleaseID=103499.

———. n.d. "Disney Designs Wild Spaces Where Man Meets Beast, Happily." Accessed February 14, 2007. http://www.wdwnews.com/ViewPressRelease .aspx?PressReleaseID=99891.

West, Paige, and James G. Carrier. 2004. "Ecotourism and Authenticity: Getting Away from It All?" *Current Anthropology 45,* no. 4:483–98.

Wetherell, Margaret, Stephanie Taylor, and Simeon Yates, eds. 2001. *Discourse Theory and Practice: A Reader.* London: Sage Publications.

Willis, Susan. 2005. "Disney's Bestiary." In *Rethinking Disney: Private Control, Public Dimensions,* ed. Mike Budd and Max H. Kirsch, 53–71. Middletown: Wesleyan University Press.

Wolfe, Cary. 2010. *What Is Posthumanism?* Minneapolis: University of Minnesota Press.

Wondrak Biel, Alice. 2006. *Do (Not) Feed the Bears: The Fitful History of Wildlife and Tourists in Yellowstone.* Lawrence: Kansas University Press.

Wright, Chris. 2006. "Natural and Social Order at Walt Disney World: The Functions and Contradictions of Civilising Nature." *Sociological Review* 54, no. 2:303–17.

WWF (World Wildlife Fund). 2007. "WWF 2007 Annual Report." Accessed February 11, 2008. http://www.nxtbook.com/nxtbooks/wwf/annualreport07/.

Yellowstone National Park. 2009. "Wolves of Yellowstone." Accessed May 25, 2011. http://www.nps.gov/yell/naturescience/wolves.htm.

———. n.d. "Wildland Fire in Yellowstone." Accessed November 28, 2007. http://www.nps.gov/yell/naturescience/wildlandfire.htm.

Žižek, Slavoj. 1997. *The Plague of Fantasies.* London: Verso, 1997.

Index

Page numbers in italics refer to photographs and illustrations.

135; ecotour through, x, xxi, 107–
34, *110*, 158, 183, 185; fire ecology
of, 119; wolves in, 121, 126, 129–
34, 136, 138
yetis, 63, 84

Žižek, Slavoj, *The Plague of Fantasies,*
  74
zoos, 74, 76. *See also* Association of
  Zoos and Aquariums (AZA)

Stephanie Rutherford is assistant professor in the Department of Environmental and Resource Studies at Trent University.